FIELD GUIDE
TO THE
PLANT LIFE
OF BRITAIN & EUROPE

KINGFISHER

FIELD GUIDE TO THE PLANT LIFE OF BRITAIN & EUROPE

GENERAL EDITOR
MICHAEL CHINERY

Kingfisher Books

ARTISTS

Russell Barnett
Norma Birgin
Wendy Bramall
Nigel Hawtin
Lindy Norton
Bernard Robinson
George Thompson

First published in this edition in 1987
by Kingfisher Books Limited
Elsley Court, 20–22 Great Titchfield Street,
London W1P 7AD
A Grisewood & Dempsey Company

BRITISH LIBRARY CATALOGUING IN PUBLICATION DATA
Kingfisher guide to the plant life of Britain
& Europe.
1. Botany – Europe 2. Botany – Great
Britain 3. Plants – Identification
I. Chinery Michael
581.94 QK281

ISBN 0-86272-212-8 Hbk
ISBN 0-86272-211-X Pbk

Design: Reg Boorer
Cover: The Pinpoint Design Company

Phototypeset by Southern Positives and Negatives (SPAN), Lingfield, Surrey
Colour separations by Newsele Litho, Milan
Printed in Italy by Vallardi Industrie Grafiche, Milan

Front Cover: Pasque Flower, Honeysuckle, Red Oak leaves and acorns, Foxy Spot, Amethyst Deceiver, Guelder Rose fruit
Back Cover: Prickly Juniper
Spine: Mediterranean Medlar, Timothy Grass

CONTENTS

INTRODUCTION

The plant kingdom is one of the two generally accepted divisions of the living world, the other being the animal kingdom. It is not usually very difficult to decide whether something is a plant or an animal. Plants are usually stationary, while animals move about, and this is clearly connected with their methods of feeding. Animals normally have to search for food, whereas most plants make their own food by the process of photosynthesis. Using the green pigment called chlorophyll, they trap sunlight and use its energy to combine water with carbon dioxide to form simple sugars. These sugars form the starting point for all the other materials made by the plant. During photosynthesis the plants give out oxygen. Plants thus play a vital role in keeping animals alive: directly or indirectly they provide everything that animals eat, and they also provide all the oxygen that animals – including the human population – need to breathe. During breathing animals give out the carbon dioxide that plants need to make their food, so there is a continuous recycling of the materials. Many scientists are concerned that the continued destruction of the world's forests, especially in the tropical regions, will upset the balance of this interdependent relationship and reduce the amount of oxygen in the atmosphere.

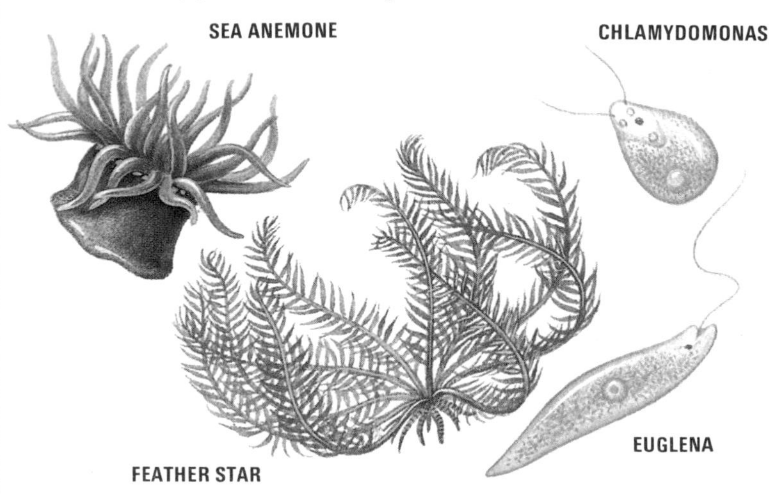

It is not immediately obvious which of these organisms are plants and which are animals. The Sea Anemone and the Feather Star are fixed to the seabed and look like plants, but they are actually animals. Chlamydomonas and Euglena are microscopic swimming plants (see page 276) with food-making chlorophyll like most other plants.

There are exceptions to the simple rule that plants are green and stationary whereas animals move about. Many of the more primitive plants live in water and swim quite actively (see page 275), and there are quite a number of animals that stay fixed in one place. Good examples of the latter include the corals and sea anemones and the barnacles. There are also some plants that have no chlorophyll and therefore have to get ready-made food from elsewhere. These include a number of parasitic plants like the Broomrape (see page 98) and all of the fungi (see pages 238-74). Some botanists consider the fungi not to be true plants at all, although some of the simpler species are structurally very similar to the algae. Their reproductive processes are also very similar.

The Vegetation Belts

The natural vegetation of Europe falls into four major zones based on climatic differences: the tundra, the coniferous belt (the taiga), the deciduous belt, and the Mediterranean forest belt. In the far north, where conditions are too cold and windy for trees to grow, is the tundra – a rather desolate plain fringing the Arctic Ocean and stretching southwards to the edge of the coniferous forests. In Europe the tundra lies entirely within the Arctic Circle and forms an arc, no more than roughly 300 km wide, around the northern coast of Scandinavia. The ground is frozen solid in winter, yet relatively little snow falls on the tundra – what does fall is often blown away by the wind. Plant life has to be extremely hardy to withstand such conditions.

During the summer, the surface layers of the tundra thaw out, but the melt water cannot drain away because the underlying strata, no more than roughly a metre down, are permanently frozen. Owing to the poor drainage the whole tundra becomes a gigantic swamp, dotted with millions of pools. Mosquitoes and midges breed in astronomical numbers on the summer tundra, and birds find it an excellent place in which to rear their families. Apart from the vast amount of insect food available to the birds, there is unbroken daylight in which to catch it. The vegetation explodes with colour as the ground thaws and becomes carpeted with brilliant flowers such as the Arctic Poppy and the Moss Campion. There are also many mosses and lichens, including the Reindeer Moss (see page 274) which forms extensive carpets up to 20 cm thick. Nothing dares to lift its head too high in this windswept habitat, however. The largest plants are dwarf birches and willows that spread very slowly over the ground. Although only a few centimetres high, their gnarled, ground-hugging

THE MAJOR VEGETATION ZONES OF EUROPE

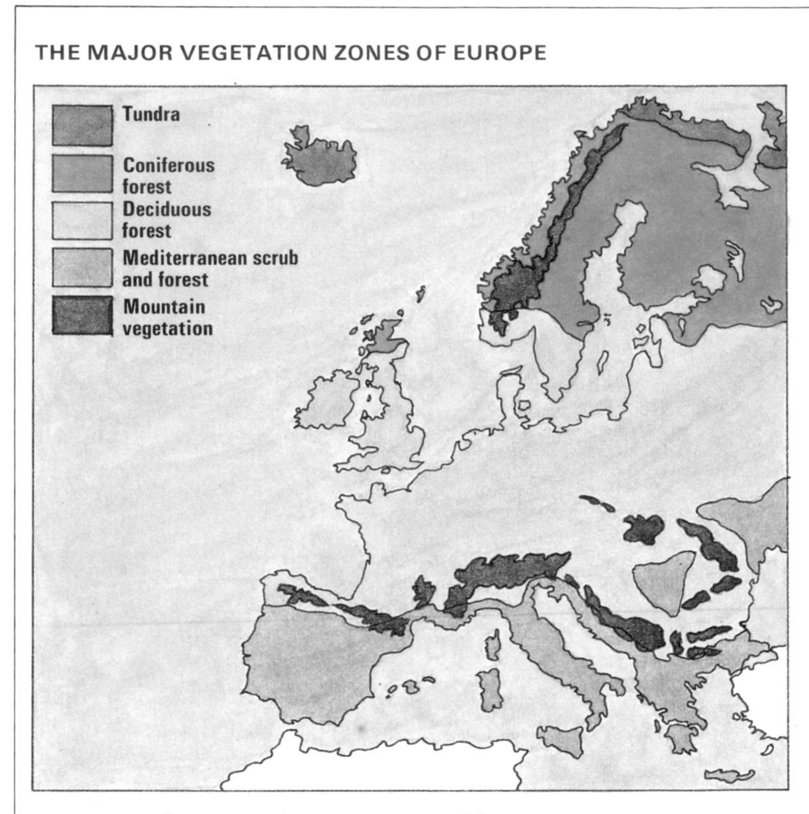

stems may be as much as a century old.

To the south of the tundra is an extensive belt of coniferous trees, often known as the taiga. It extends southwards almost to the southern tip of Sweden and is composed mainly of Scots Pine and Norway Spruce. It is largely evergreen, although birches and willows grow in profusion along its northern borders. Rainfall is much higher in the taiga than on the tundra – which is almost a desert in many places – but the winters are still very cold. The leaves of the evergreen are well suited to these conditions. They have thick waxy coats which cut water loss to a minimum and so prevent the trees from drying out in winter, when they cannot absorb water from the frozen soil. They are, however, green and ready to make food as soon as the temperature rises in spring.

Northern Scotland is just included in the coniferous belt, although only a few small patches of the original pine forest remain today. Many of our coniferous woodlands, even in Scandinavia, are now man-made plantations.

Most of Europe south of the Baltic Sea, including almost all of the British Isles, lies in the deciduous forest belt. Were it not for human activity, the whole area would still be clothed with oaks and other broad-leaved trees. The cool or cold winters, together with warm summers and rainfall distributed fairly evenly throughout the year, provide perfect conditions for the growth of deciduous trees – but the climate is also ideal for agriculture and vast tracts of the original forest cover have been removed. Most of the deciduous forest belt is now occupied by grassland, though it quickly reverts to woodland if grazing or cultivation ceases. Some areas have already reverted, and some have been deliberately replanted with trees, but these secondary woodlands are never as rich in plant and animal life as the truly ancient forest remnants.

The last major vegetation zone is the Mediterranean forest; composed largely of evergreen trees, it occupies the hottest parts of Europe close to the Mediterranean Sea. This forest contains both conifers and broad-leaved evergreens such as the Cork Oak (see page 175). Growth takes place mainly in the rather mild winter, when most of the rain falls, and the thick waxy leaves do not lose too much water during the long summer drought. However, as with the deciduous forests further north, 8,000 years of human activity have destroyed most of the original vegetation of the Mediterranean belt. Today there are many barren hillsides and the rest of the area is covered largely with *maquis* and *garrigue* – dense scrubby areas dominated by the cistuses (see page 214) and other low-growing shrubs.

Within each of the four zones there are many variations in soil and drainage, dependent on the underlying rocks, and these have a marked effect on the types of plants growing in each area. Within the deciduous forest zone, for example, there are ashwoods, beechwoods and various kinds of oakwoods. Each zone also has its freshwater habitats, its cliffs and seashores, its heaths and moors, and, of course, its mountains. Altitude also plays a major part in the distribution of plant life – the mountains support a vegetation totally different from that of the neighbouring lowlands, even when in the same zone. Since the vegetation changes with both altitude and temperature, climbing in the southern Alps can be rather like journeying from the Mediterranean to the tundra. The journey starts amid the tough-leaved Mediterranean forest or scrub, and then passes through the deciduous and evergreen coniferous zones before reaching the ground-hugging alpines which are very similar to the plants of the tundra. In fact, many species are found in both habitats.

Each of the habitats mentioned above has its own assemblage

of plant life. Heathland, for example, is dominated by heathers and a few other shrubs, while coastal salt-marshes are populated largely by members of the goosefoot family (see page 44). Grassland on alkaline soils carries an assemblage of grasses and other plants rather different from that found on acidic soils. Even the smaller plants of the oakwood tend to be different from those growing in the ashwood. These variations are all connected with soil, moisture and shade. Beechwoods, for example, cast such a dense shade that very few small plants can survive under the trees at all. The habitat in which a plant grows can often be a useful clue to its identification. The Water Forget-me-not (see page 89), for example, is readily distinguished from the rather similar field and wood forget-me-nots simply by its liking for marshy habitats.

The study of plant associations is a very interesting branch of botany and one which the amateur can follow very easily. By systematically recording which plants grow in each of the various habitats you explore, you can soon get to know the likes and dislikes of many species. An experienced botanist can often predict many of the plants likely to be found in an area just by looking at it from a distance. Many plant associations obviously occur since species enjoy the same physical conditions, but there are also examples of plants which like, or even need, each other's company. The Fly Agaric fungus (see page 240), for example, has a strong liking for birch trees and its underground threads actually enter into a mutually beneficial relationship with the birch roots. Such partnerships between fungi and other plants are described as mycorrhizal and they are actually extremely common. Most woodland toadstools, for instance, probably enter into such relationships with at least one or more tree species.

This book will help you to identify many of the plants that you come across in the various habitats throughout Europe. Obviously it can cover only a small proportion of the thousands of species that grow in the region, but most of the common flowering plants and conifers are included as are most of the larger and more conspicuous fungi. Unless otherwise stated, it may be assumed that the plants occur in suitable habitats throughout the region. The flowering times given are usual for northern and central Europe, but it should be remembered that the same species may flower much earlier in southern Europe.

The Classification of Plants

The plant kingdom has been divided up in many ways in the past and there is still no universally accepted system of classification. The chart given here shows the main groups of living plants and their relationships to each other. It must be pointed out, however, that many different scientific names are used for these same groups of plants. One modern scheme, for example, uses Arthrophyta (jointed plants) for the horsetails and Anthophyta for the flowering plants. But names *are* only names and they do not alter the nine main groups into which our plants are traditionally divided.

THE MAIN GROUPS OF PLANTS

THALLOPHYTES	ALGAE	Divided into several distinct groups according to the nature of their pigments
	FUNGI	Totally lacking chlorophyll and divided into several groups according to their reproductive structures and processes
CORMOPHYTES	BRYOPHYTES*	Mosses
		Liverworts
	PTERIDOPHYTES	Ferns
		Clubmosses
		Horsetails
	SPERMATOPHYTES	Gymnosperms – the conifers and their allies
		Angiosperms – the flowering plants

*Some modern classifications use the name Bryophyte just for the Mosses and refer to the Liverworts as Hepatophytes.

The Thallophytes are simple, flowerless plants without any division into root, stem and leaf. The body is known as a thallus and food materials are absorbed all over its surface. Within this group there are two main sections: the Algae and the Fungi. The Algae are a group of some 20,000 species, nearly all of which live in water. They range from microscopic single-celled organisms to massive seaweeds several metres long. Although they all contain chlorophyll and can make their own food, they also contain numerous other pigments. The Algae are placed in several distinct groups according to the nature of these pigments, the various groups being only distantly related. The green algae (see page 276) are believed to have been the ancestors of the first land plants.

The Fungi (see pages 235-74) are chemically unlike the other plants. They have no chlorophyll, and therefore cannot make food in the normal way, and their bodies are composed of a type of cellulose not found elsewhere in the plant kingdom. For these reasons some biologists do not consider the fungi to be plants. They are, however, structurally quite similar to some of the simpler green algae and they also reproduce in much the same way as these algae. The 50,000 or so known fungi species are placed in several distinct groups, according to their methods of reproduction (see pages 235-7). Some of the groups, notably the slime moulds (not included in this book), are only distantly related to the familiar mushrooms and toadstools.

The Thallophytes also include the microscopic, single-celled Bacteria, although these cannot be said to have anything like a thallus. They are closely related to the blue-green algae and it is likely that the earliest of all living things belonged to these two groups.

The rest of the plant kingdom belongs to a group variously known as the Cormophytes or the Embryophytes. With the exception of the thalloid liverworts (see page 234), the Cormophytes typically have an aerial shoot with leaves – although the latter may be much reduced in some species – and an underground root system which absorbs water and minerals.

Within the Cormophytes there are three main sections: the Bryophytes, the Pteridophytes and the Spermatophytes. The Bryophytes (see pages 232-4) include the mosses and the liverworts – all rather small green plants with very simple roots. Most are confined to damp places and they reproduce by scattering tiny spores from rather prominent capsules (see page 232). There are about 15,000 known species of mosses and nearly 6,000 different liverworts, the majority living in tropical regions.

The Pteridophytes (see pages 226-31) include the ferns and their relatives, green plants which are mostly somewhat larger than the mosses and liverworts. Unlike the latter, the ferns and their relatives have a good root system and an elaborate arrangement of tubes for carrying water and food materials through the plant. The plants have no flowers but carry spores on various parts of the body. The ferns – variously known as Filicales, Filicopsida, Pteropsida and Pterophyta – usually have large, finely-divided leaves or fronds. The spore capsules are usually on the undersides of the fronds (see page 227). The plants are mostly terrestrial, although a few aberrant species live in water (see page 228).

The clubmosses, also belonging to the Pteridophytes, have small leaves on trailing or upright stems and can easily be mistaken for mosses until they produce their spore capsules, which are commonly clustered into club-shaped cones (see page 230). The group, with about 1,000 living species, is mainly tropical and is technically known as the Lycopsida.

The horsetails (see page 231) are a group of Pteridophytes in which the leaves are reduced to collars of tiny scales around the slender green stems. The latter may or may not bear whorls of slender branches, but they are always distinctly jointed – hence the name Arthrophyta which is sometimes used for the group instead of the older name Sphenopsida. The Great Horsetail, Europe's largest species, can occasionally reach a height of about two metres, but most horsetails are much smaller than this – although their prehistoric ancestors topped 30 m in the ancient coal forests. Their spores are carried in soft cones at the tops of the stems.

The Spermatophytes are the seed-bearing plants, of which the seed is a reproductive body consisting of an embryo plant and a supply of food inside a tough protective coat. This is a great advance on the reproductive processes of the spore bearing ferns and mosses (see page 227), for it does not necessitate the existence of free water at any time. The two main divisions within this section are the Gymnosperms, which include the conifers and their relatives, and the Angiosperms or flowering plants.

Gymnosperm means 'naked seed', and most of the Gymnosperms actually bear their seeds more or less nakedly on the scales of woody cones. One or two species, including the Yew (see page 164), carry their seeds singly and partly surrounded by a fleshy cup, but the pollen is produced in very definite cones and there is no doubt that these plants belong with the conifers. The leaves are generally narrow and often needle-like, and the plants are all trees or shrubs (see pages 144-7).

Angiosperm means 'cased seed' and refers to the fact that the flowering plants all enclose their seeds in fruits, although the latter commonly split open to release the seeds when ripe. There are two major groups of flowering plants – the Dicotyledons and the Monocotyledons (see page 23).

The Thallophytes and the Cormophytes are sometimes regarded as subkingdoms of the plant world, but this suggests that they are of equal rank when clearly they are not. As we have already seen, many of the groups of algae are very different from each other and, in biological terms, they may be just as distantly

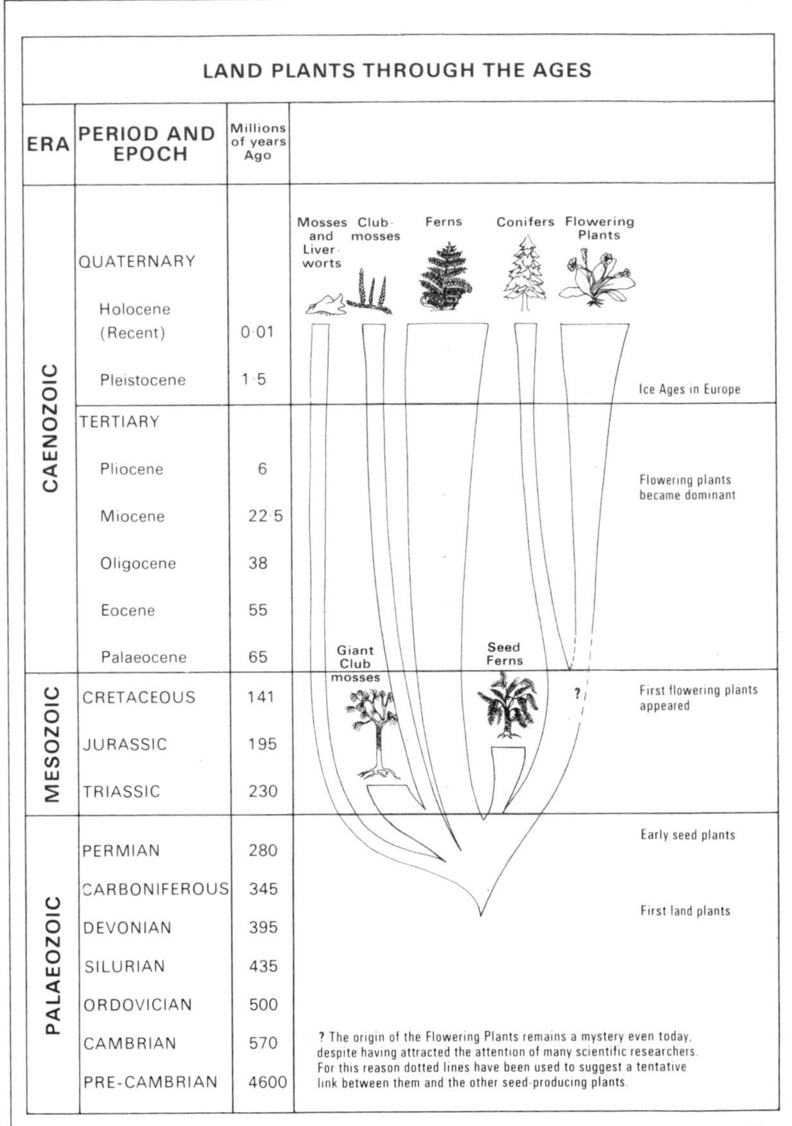

related to each other as they are to the Cormophytes – especially in view of the generally accepted belief that the green algae gave rise to the land plants and thus to all the Cormophytes. The terms Thallophyte and Cormophyte are therefore merely convenient labels and, although occurring in many books, they have little real value in classification.

Plant Families and Genera

All the plant groups just described have a number of subdivisions, of which the most important are the families. The members of a family have a good deal in common, and among the flowering plants, it is the flower itself that is of greatest value in classification and identification. Flower structure remains more or less constant within a family – although there are some exceptions – and you can very often place a plant in its family with just one glance at a flower. You are then well on the way to identifying the species. A good example is the pea family, with its characteristic hooded flowers (see pages 48-54). Leaves and stems vary considerably according to the life form of the species and also according to the conditions in which it is growing, so cannot be relied on to separate the families. Trees, shrubs and low-growing herbs can all be found in the same family, a good example of this variety being the rose family (see pages 55-61) which contains apples and strawberries as well as the roses. Family names almost all end in -aceae.

The families are further divided into genera (singular: genus), each of which contains one or more individual kind or species of plant. The members of a genus are all extremely closely related, although they may have very different growth forms. The Water Crowfoot (see page 26), for example, belongs to the same genus as the Meadow Buttercup (see page 25), although their leaves are totally different as a result of their completely different life styles. The flowers differ in colour, although structurally they are alike.

All the flowering plants and conifers in this book are assigned to their families, and most families are given a short introduction in which their main features are pointed out. Most species are given both English and scientific names, the latter consisting of two parts and conventionally printed in italics. The first part is the generic name (the name of the genus) and is shared by all the members of the genus, while the second part is the specific name. Together, the two parts positively identify one particular kind or species of plant and are understood by botanists all over the world. Some of the fungi and other smaller plants have never received English names, so only a scientific name can be given.

Studying Plants

In the past botanists sallied forth with trowels and collecting boxes and returned home with all kinds of plant material to be pressed or otherwise dried for their collections. Luckily for our wild plants, the camera has now for most people taken the place of the trowel and vasculum. You will do no harm by picking a few common flowers, but remember that it is the flowers that

The Large Yellow Gentian (centre), the Martagon Lily (left) and the Wild Gladiolus all grow wild in various parts of Europe. Although common in some areas, they are legally protected in others. Only the Wild Gladiolus occurs in Britain, where it is rare and strictly protected.

produce the seeds for the next generation and that if too many flowers are picked the population is bound to suffer in the long term. It is illegal to dig up wild plants in Britain without the landowner's permission, and there are some real rarities, such as the Military Orchid, the Wild Gladiolus and the Spring Gentian (page 88), that must not even be picked. Similar regulations are in force in many other parts of Europe, especially in mountainous areas where many alpine plants are threatened because of their popularity as garden rock plants.

Always keep your notebook and pocket lens handy. Make your notes and sketches while you are looking at the plant, and remember to include leaves from the base as well as from higher on the stem in your sketches. Take your notebook and field guide to the plant: don't pick the plant and take it to the book.

Unless you are interested in photography, you need very little equipment for studying the wild flowers and other plants around you. Your field guide is, of course, indispensable, as is your notebook. Plants do not wander away before you can get out your field guide and name them, so you don't have to be able to make lightning sketches like the birdwatcher, though it is useful to make sketches of the less familiar plants you find. Such drawings can be of great help in recognizing the plants the next time you come across them. Don't forget to include some idea of scale, and sketch in leaves as well as flowers. A leaf pressed between the pages of your book will serve as a permanent reminder and picking just one leaf is unlikely to harm a mature plant. Record the time of flowering and fruiting, the type of soil, and the altitude and aspect of the area; also make a note of the surrounding plant species, and you will then build up a thorough knowledge of the vegetation of a particular area.

A good pocket lens is essential for looking at mosses and other small plants, and for examining flowers in detail. A folding lens is generally the most useful and it should have a magnification of between × 8 and × 15, although the higher magnification will give you very little working distance. Binoculars can be useful in some circumstances – for examining plants growing on cliffs and in the middle of ponds, for example. They are also useful for scanning trees and looking at their flowers or at lichens or other plants growing high on their trunks or branches. A very good specification for botanical binoculars is 8 × 30, magnifying eight times and yet not very heavy. Don't be tempted to go for much higher magnifications, for they will not allow you to focus on plants which are only just out of reach.

Some Simple Rules

Botanists and photographers have combined to draw up a set of rules designed to safeguard our wild plants. They can be summed up as follows:

- Always seek the permission of the landowner if you want to look for flowers on private land.
- Take your field guide to the plant, not the other way round. If this is not possible, take only the smallest adequate piece of the plant away for identification – and don't take anything other than photographs if you think the plant may be a rarity.

- Pick only flowers which are common in the locality; it is better not to pick any and to leave them for others to enjoy.
- Don't tell people where to find plants that you think are rare, although you should inform your local Naturalists' Trust so that they can make a check and take measures to safeguard plants if necessary.
- Don't do anything which would expose a rare plant to unwelcome attention. A trampled area around a plant can make it very obvious; the trampling can also damage surrounding plants. Always restore an area to as natural a state as possible after looking at or photographing plants and give no one cause to regret your visit.
- Photographers should remember that the welfare of the subject is more important than the photograph.

A sensible attitude to the plant life around us will ensure that it will still be there for the enjoyment of future generations.

The Country Code

1. Leave no litter.
2. Fasten all gates.
3. Avoid damaging fences, hedges and walls.
4. Guard against all fire risk.
5. Keep dogs under control.
6. Keep to paths across farmland.
7. Safeguard all water supplies.
8. Protect all forms of wildlife.
9. Go carefully on country roads.
10. Respect the life of the countryside.

CHAPTER ONE

WILD FLOWERS

The flowering plants we see around us, both in our gardens and in the wild, belong to the Angiosperms, the most advanced group of plants. The Angiosperms are found in every kind of habitat – from the highest mountain to the seashore, from the frozen tundra to the hottest desert. Many live in water, although only a handful of the 250,000 or so known species have managed to invade the sea. Flowering plants range from minute floating duckweeds, no larger than a pin-head, to mighty trees which tower 100 metres or so high. This chapter, however, is concerned mainly with the herbaceous soft-stemmed flowers of the countryside.

Flowers

The flower is the reproductive part of the plant, where the seeds are formed ready for the next generation. A typical flower, such as that shown here, consists of several greenish sepals on the outside, a number of colourful petals, several or many pollen-producing stamens and one or more carpels, which eventually form the fruits and enclose the seeds. The upper surface of the carpel is known as the stigma. The sepals are collectively known as the calyx and the petals form the corolla: together they are known as the perianth.

The number and arrangement of the various parts of the flower are of great importance in classification and identification. When trying to identify a flower always start by counting the sepals and petals and taking note of whether or not they are joined together to form tubes or bells. It will also be necessary to decide whether the carpels – collectively known as the gynaecium or ovary – are above or below the attachment of the petals.

Many books, especially those of a more technical nature, describe the basic structure of a flower by giving its floral formula. The Meadow Cranesbill, for example, would be described as follows: K5 C5 A10 $G(\underline{5})$. This tells us that the flower has five sepals (K=calyx), five petals (C=corolla), ten stamens (A=androecium, the collective name for the stamens) and five carpels (G=gynaecium). The brackets after the G indicates that the 5 carpels are joined together; the line under the 5 indicates that the carpels are above the attachment of the petals. A formula such as this is of great use to the botanist; it often indicates the plant's family immediately – although it provides no

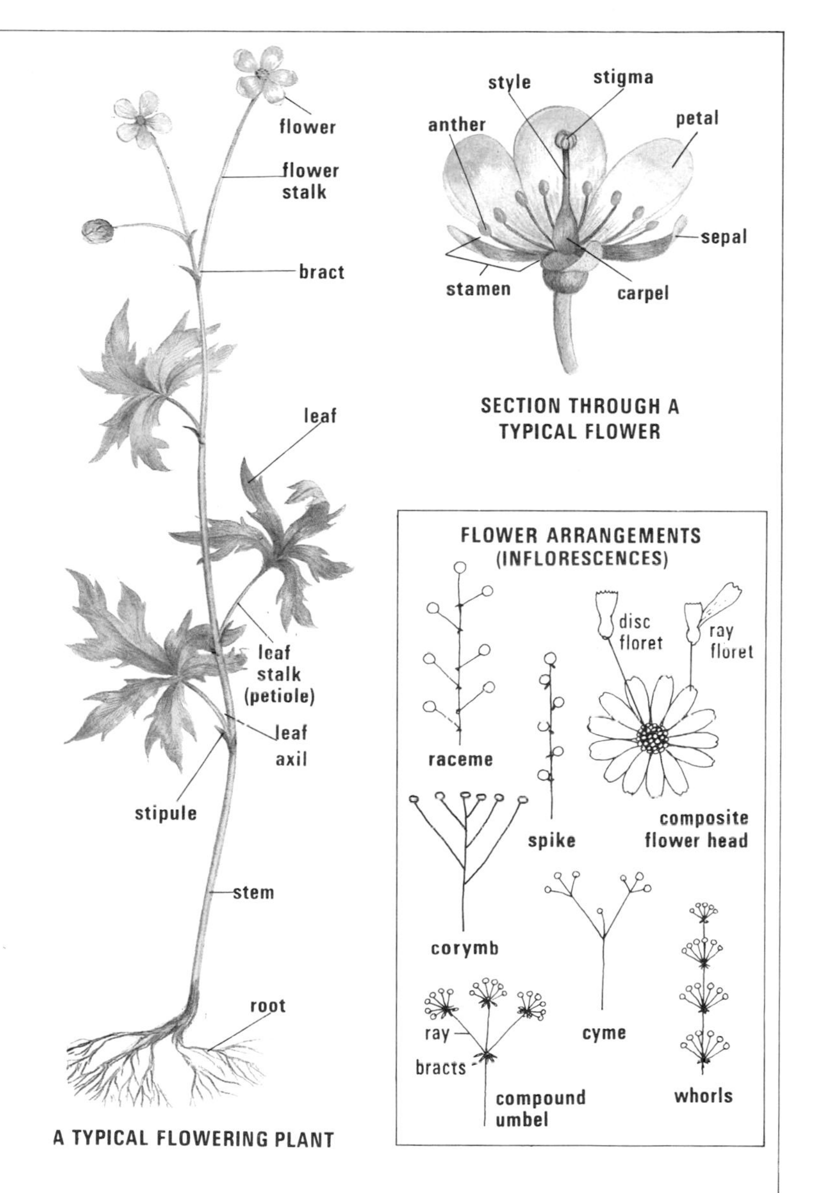

A TYPICAL FLOWERING PLANT

SECTION THROUGH A TYPICAL FLOWER

information about the colour or shape of the flower. The flowers of many plant families have very characteristic shapes; learn to recognize the typical shapes and flower identification will become much easier.

Inflorescences

Flowers may be carried singly on their stems, as in tulips and wood anemones, but they are much more often grouped into clusters called inflorescences. Some of the main types of inflorescences are shown in the diagram. In a cyme, the main growing point ends in a flower and growth continues by one or more side branches until they too end in flowers; the oldest flower is thus in the centre of the inflorescence. In the other types of inflorescence, the main axis goes on growing until the flower head is complete; the oldest flowers are then at the bottom or the outside. The commonest inflorescence of this type is the raceme: flowers appear at intervals on one or both sides of the main stem, as in the foxglove. A spike is a raceme in which the individual flowers have no stalks; a panicle is a branched raceme. A corymb is a type of raceme in which the lower flower stalks are longer than the upper ones, thus bringing all the flowers to about the same level, as in the Yarrow (see page 122). An umbel is superficially similar to a corymb, but in the umbel all the flower stalks spring from the same point on the stem. This type of inflorescence is characteristic of the carrot family (see pages 68-74). Daisies and other members of the family Compositae (see pages 111-23) have masses of very tiny flowers called florets, packed tightly into the flower head.

Seeds and Fruits

Seeds cannot develop until pollen of the right type falls on to the stigma on top of the carpels. This process of pollination may be brought about by the wind or by the activities of insects. Brightly coloured petals attract insects, but wind-pollinated flowers often have small and dull petals, or none at all. In some plants either carpels or stamens are missing, giving rise to male or female flowers respectively, but the male flowers cannot, of course, produce seeds.

After pollination, cells from the pollen and the carpels fuse to initiate the seed; the carpels then swell up to form the fruits. The name Angiosperm actually means 'cased seed' and refers to the fact that the seeds are encased in the fruits. Fruits are of many kinds: some are quite dry when ripe while others are very juicy. Many dry fruits split open to release their seeds when they are ripe. Of the dry fruits, a follicle is a slender fruit that splits down one side, while a pod – characteristic of the pea family – splits down both sides. Capsules vary in shape and open in various ways. An achene is a small dry or leathery fruit with one seed. It does not split open and the germinating seed has to push its way out. Berries are juicy fruits with many seeds, the latter being scattered by birds which eat the juicy flesh.

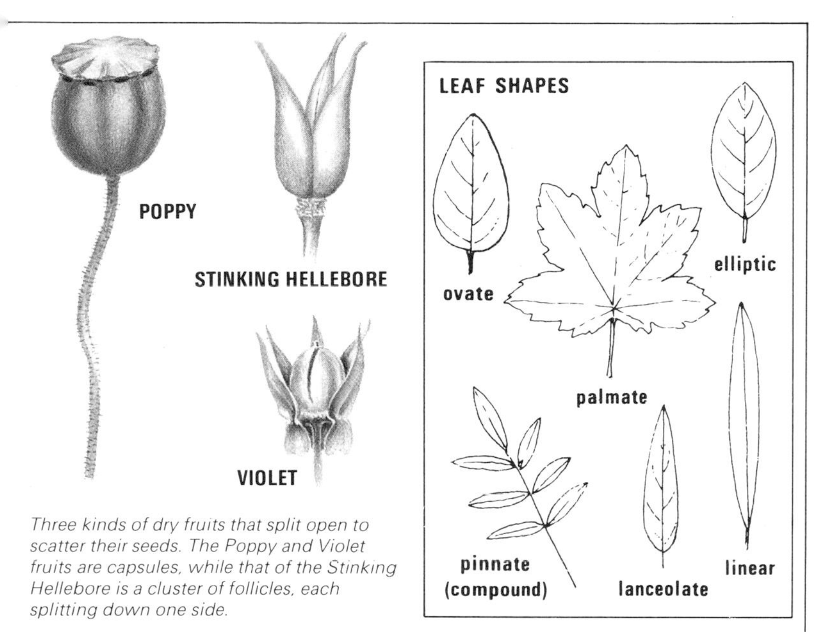

Three kinds of dry fruits that split open to scatter their seeds. The Poppy and Violet fruits are capsules, while that of the Stinking Hellebore is a cluster of follicles, each splitting down one side.

Leaves

Leaf shapes can be important in identifying plants; several common shapes are shown in the diagram. The margin of a leaf may be quite smooth (in which case it is called entire), but it is more often toothed or lobed. Compound leaves are split into several distinct leaflets. Bracts are small leaves at the bases of flower stalks, while stipules are outgrowth from the bases of leaves. The leaf stalk is called the petiole.

The flowering plants are split into two major divisions: the Dicotyledons, in which the leaves are relatively broad, and the Monocotyledons, in which the leaves are generally strap-shaped. The main difference between the two groups lies in their seeds: dicotyledon seeds have two leaves inside them, while monocotyledon seeds have only one. Each group is then split into numerous families, the divisions being based mainly on the structure of the flower. Leaves and stems vary greatly according to the conditions under which the plants are growing so they cannot always be relied on for classification. Family names nearly all end in -aceae.

Unless otherwise stated, all the plants described in this chapter are perennials – plants which live for several or many years. Annuals complete their life cycles in one year or less, while biennials live for two years.

Dicotyledons

BUTTERCUP FAMILY Ranunculaceae

Most of the 2,000 or so plants in this family are herbaceous species from the north temperate regions. The flowers are regarded as rather primitive, with few specializations. There are many stamens and usually many carpels. The sepals, normally 5 in number, are often petal-like and true petals are then absent. When present, there are five separate petals.

WOOD ANEMONE
Anemone nemorosa

A rather delicate, hairless plant of broad-leaved woodland, hedgebank and upland meadow, it grows 6-30 cm high. Like many woodland plants the Wood Anemone makes use of the sunlight of early spring for flowering before the trees come into leaf.
Flower: white or pink-tinged, 2-4 cm across, no petals but 6-7 petal-like sepals; stamens many.
Flower arrangement: solitary, terminal.
Flowering time: March-May.
Leaf: 1 or 2 long-stalked basal leaves of 3 segments, each segment deeply dissected, coarsely toothed, shortly stalked; upper leaves in whorl of 3, smaller with flattened stalks.
Fruit: many downy achenes.

MARSH MARIGOLD
Caltha palustris

Although this stout plant has flowers that look like large buttercups, its broad, shining, heart-shaped leaves and hollow flower stalks betray its true identity. It grows in marshy ground, often near streams, and may either be erect, to a height of 15-30 cm, or creeping.
Flower: bright golden yellow, 1·5-5 cm across; petal-like sepals, 5 or more; stamens many.
Flowerarrangement: few-flowered cyme.
Flowering time: March-July.
Leaf: broadly heart-shaped, teeth blunt or pointed; basal leaves long-stalked, upper often kidney-shaped, stalkless.
Fruit: head of 5-15 follicles, erect or curved back.

MEADOW BUTTERCUP
Ranunculus acris

Commonly found in meadows and pasture, this buttercup grows up to 100 cm high. The stem is much branched and erect above.
Flower: glossy, bright yellow, 1·5-2·5 cm across; petals 5; sepals lying against petals; stamens many; flower stalk not furrowed.
Flower arrangement: irregular cyme.
Flowering time: April-October.
Leaf: more or less hairy; lower leaf with long stalk, palmate with 2-7 deeply toothed segments, middle segment not stalked; uppermost leaves stalkless, deeply cut.
Fruit: achenes many, hairless, each with short hook, in a round head.

BULBOUS BUTTERCUP
Ranunculus bulbosus

This plant, named for the bulb-like tuber at the base, resembles the last species but grows in drier grassland. The middle leaf-lobe has a long stalk, and the flower-stalk is furrowed. The stem is less branched, but when in flower the plant is best distinguished by the reflexed sepals. It grows to about 50 cm, flowering April-July.
The very similar Creeping Buttercup (*R. repens*), a common garden weed, has a creeping, rooting stem. Its middle leaf-lobe is stalked and the flower stalk is furrowed.

LESSER CELANDINE
Ranunculus ficaria

One of the early spring flowers, Lesser Celandine has more petals than a buttercup and glossy, heart-shaped leaves. It grows 5-30 cm high on hedgebanks, in woods and stream-sides. The roots form tubers.
Flower: bright golden yellow, whitening with age at petal-base, opening only on fine days, 1·5-5 cm across; petals 8-12, more or less pointed; sepals 3; stamens many.
Flower arrangement: solitary.
Flowering time: March-May.
Leaf: broadly heart-shaped, dark green, hairless; basal leaves long-stalked, in a rosette.
Fruit: achenes, many, downy.

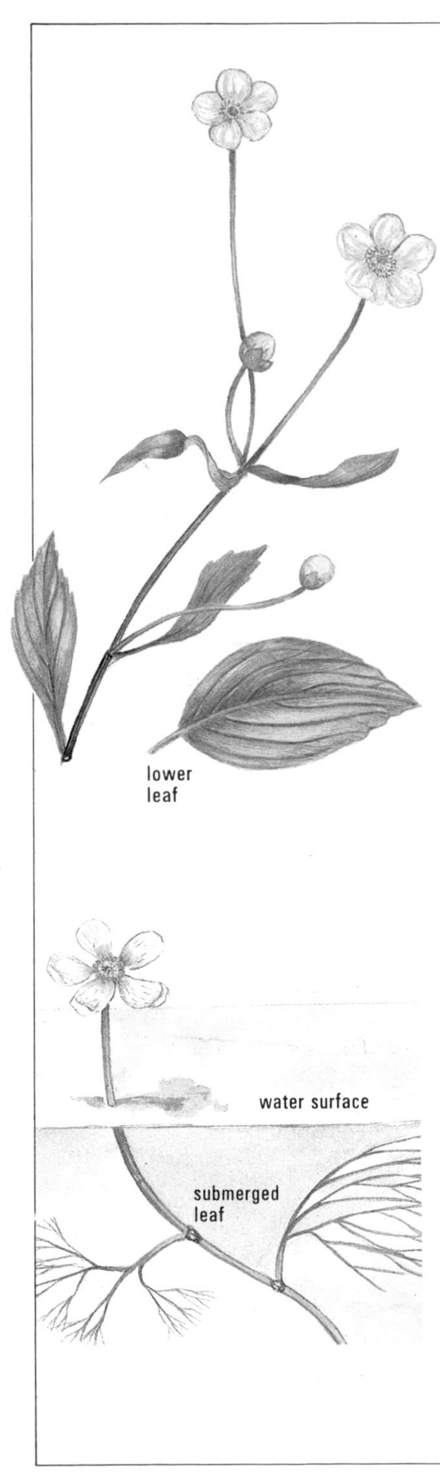

LESSER SPEARWORT
Ranunculus flammula

This is a Buttercup of wet places easily distinguished by its undissected leaves and pale, rather than golden, yellow flowers. Erect or creeping, with a hollow stem, and usually hairless, it grows between 8 and 80 cm high. Irritant poisons are present in larger amounts than in other buttercups.

Flower: pale yellow, glossy, 8-20 mm across, petals 5; sepals 5; stamens many; flower stalk grooved.
Flower arrangement: solitary or in a cyme.
Flowering time: May-September.
Leaf: lower leaves ovate, stalked; upper leaves stalkless, smaller, pointed, parallel-veined, with or without teeth.
Fruit: achenes, hairless, in a round head.

THREAD-LEAVED WATER-CROWFOOT
Ranunculus trichophyllus

This Water-crowfoot has small, white flowers and grows in ponds, slow streams and ditches. There are no floating leaves, and the needle-like segments of the submerged leaves do not lie flat. Several other species of Water-crowfoot grow in N. Europe, some of which have floating leaves in addition to submerged.

Flower: white, petal-base yellow, 8-10 mm across; petals 5, not touching at edges; stamens 5-15.
Flower arrangement: solitary, opening above surface of water.
Flowering time: May-June.
Leaf: all submerged, finely dissected, bristle-like, not lying in one plane, 2-4 cm, shortly stalked.
Fruit: hairy achenes; stalk usually less than 4 cm.

PYRENEAN BUTTERCUP
Ranunculus pyrenaeus

A low-growing mountain species of damp limestone pastures in the Alps as well as in the Pyrenees, it occurs at altitudes from 1,700 to 2,800 m.
Flower: white, 10-20 mm across with 5 rounded petals and whitish sepals: many yellow stamens.
Flower arrangement: solitary or in twos and threes. May-July.
Leaf: bluish-green, hairless: strap-shaped, untoothed and unstalked (sometimes broader in Alps).
Fruit: a cluster of achenes.

GLOBE FLOWER
Trollius europaeus

Named for the almost spherical shape of the flower, this plant grows in damp woods and meadows, mainly in upland regions but also lower down in the north. It reaches 60 cm.
Flower: golden yellow, 2·5-5 cm across: 5-15 petal-like sepals forming the globe and enclosing the strap-shaped nectar-secreting petals and numerous stamens and carpels.
Flower arrangement: usually solitary: May-August.
Leaf: shiny, deep green above and paler below: lower ones stalked and deeply 3-5 lobed, upper ones stalkless and 3-lobed.
Fruit: a cluster of beaked follicles.

TRAVELLER'S JOY
Clematis vitalba

Also known as Old Man's Beard, this climber – one of the few woody plants in the family – can reach 30 m, scrambling over shrubs and trees and clinging to them with its twining leaf stalks. Found from the Netherlands southwards, but only on chalk and limestone in northern parts of its range.
Flower: 2 cm across, with 4 greenish-white sepals and no petals: fragrant.
Flower arrangement: dense panicles. June-August.
Leaf: pale green: compound, with 3-5 well separated heart-shaped or oval leaflets, each up to 10 cm long.
Fruit: clusters of hairy achenes, each with a long plume – the 'beard'.

PASQUE FLOWER
Pulsatilla vulgaris

This hairy plant, with feathery leaves, grows in chalk and limestone grassland, usually on south-facing slopes: widely distributed in northern and central Europe, but generally uncommon. It reaches 30 cm.
Flower: 6 sepals, deep purple above, paler and silky below: 5-8 cm across. Numerous stamens and carpels. Upright at first and then nodding. A whorl of hairy bracts below flower.
Flower arrangement: Solitary. April-June.
Leaf: all from base: finely divided and coated with grey hairs.
Fruit: a cluster of achenes, each with a feathery plume.

STINKING HELLEBORE
Helleborus foetidus

This much-branched evergreen plant, named for its unpleasant smell, is widely distributed in the southern half of Europe, where it grows in open woods and scrub on limestone slopes. It reaches 80 cm.
Flower: shaped like an inverted cup, up to 3 cm across. Petals absent: 5 yellowish-green sepals, edged with purple: many stamens and usually 3 carpels.
Flower arrangement: a drooping cyme. January-April.
Leaf: all on stems (none radical): lower ones dark bluish-green, palmately divided with up to 11 narrow leaflets: upper ones paler and undivided.
Fruit: a cluster of beaked follicles surrounded by opened sepals.

POPPY FAMILY Papaveraceae

The flowers of this family have 2 sepals, which fall as the flowers open, and 4 petals – often very crinkly in the poppies themselves. The plants all contain a white or yellowish latex. Several kinds of poppy are weeds of cultivation, although less common today owing to the use of herbicides. The weed species are mostly annuals, with long-lived seeds that can germinate after being buried for many years.

GREATER CELANDINE
Chelidonium majus

There seems little resemblance between this plant, with its small, yellow flowers, and its flamboyant relatives, the Poppies. Despite the common name, it is not related to Lesser Celandine. It has surprisingly bright-orange sap and grows 30-90 cm high on hedgebanks and walls, usually near buildings.
Flower: bright yellow, 2-2·5 cm across; petals 4; stamens many.
Flower arrangement: umbel of 2-6 flowers.
Flowering time: May-August.
Leaf: pinnate, leaflets 5-7, glaucous, blunt-toothed.
Fruit: a capsule 3-5 cm, straight, opening from below by 2 valves.

LONG-HEADED POPPY
Papaver dubium

The flowers are smaller and more orange than those of the Common Poppy. The stem is stiffly hairy. The plant grows 20-60 cm high, in waste places and as a weed of cultivation.
Flower: orange-pink, sometimes with a dark spot at petal-base; 3-7 cm across; petals circular, overlapping at base; stamens many, purple; stigma disc flat with 7-9 radiating ridges.
Flower arrangement: solitary.
Flowering time: June-July.
Leaf: greyish green, shortly hairy; basal leaves deeply lobed, end lobe not enlarged.
Fruit: long, narrow capsule, 1·5-2 × 0·5-0·75 cm, tapered at base, hairless, often ribbed. Opening by pores under the cap.

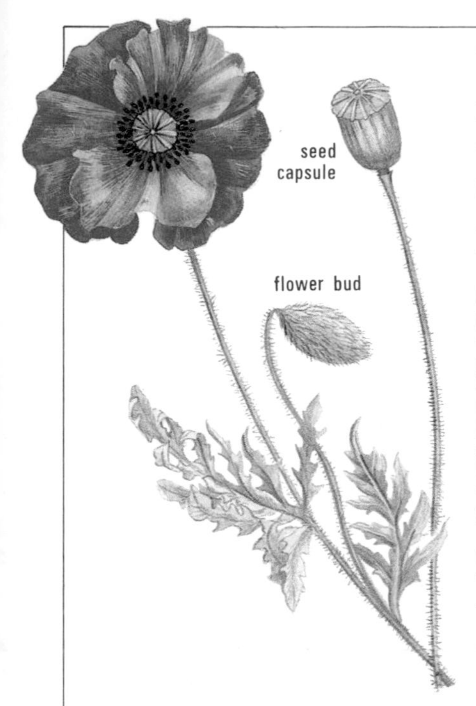

COMMON POPPY
Papaver rhoeas

The Common Poppy, with its large, scarlet flowers and covering of stiff hairs, grows 25-90 cm high. If cut, it exudes milky-white latex. Like the previous species, it is an annual and a very common plant of roadsides and field margins.
Flower: scarlet, often with dark blotch at petal-base; 7-10 cm across; petals thin; stamens many; stigma disc more or less flat with 8-12 radiating ridges.
Flower arrangement: solitary.
Flowering time: June-August.
Leaf: stiffly hairy, 1- to 2-pinnate; segments coarsely toothed.
Fruit: capsule, rounded at base, top flat, length 1-2 cm, hairless, opening by pores under top.
Uses: Poppies were thought to cause headaches and thunderstorms, but were also used to treat headaches. Although slightly poisonous, the red poppy does not contain the opium of the Opium Poppy (*P. somniferum*).

WATER-LILY FAMILY Nymphaeaceae

Water-lilies grow in still or slow-moving fresh water, with their roots and stout rhizomes in the mud. Leaves and flowers grow up on long stalks, most leaves floating on the surface and often covering large areas. The flowers may float or stand clear of the water. Like the buttercups, the water-lilies have flowers with many stamens and no specialized pollination mechanisms.

WHITE WATER-LILY
Nymphaea alba

The White Water-lily grows in standing or slow-flowing water up to a depth of 3 m. It is more tolerant of polluted water than many other water plants. The large, floating flowers are scented.
Flower: white, 10-20 cm across, floating; 20-25 pointed petals; sepals 4; stamens many; 15-20 radiating stigmas.
Flower arrangement: solitary, on long stalks.
Flowering time: June-September.
Leaf: floating, more or less circular, 10-30 cm, entire with deep cleft where stalk joins, veins joined around edge of leaf.
Fruit: flask-shaped, sometimes spherical, ripening underwater; seeds float, aiding dispersal.
Uses: the rhizomes were sometimes eaten in parts of N. Europe, and Elizabethans ate seed and root to promote chastity.

YELLOW WATER-LILY
Nuphar lutea

Growing in freshwater lakes, ponds and slow streams, the Yellow Water-lily is more widely distributed than the White. The floating leaves are oval compared with the almost circular White Water-lily leaves.
Flower: bright yellow, cup-shaped, smelling of alcohol, 4-6 cm across, rising clear of the water; sepals 4-6, large, petal-like, rounded; stamens many; 15-20 radiating stigmas on top of flask-shaped ovary.
Flower arrangement: solitary.
Flowering time: June-August.
Leaf: floating, oval, 12-40 × 9-30 cm, entire with a deep cleft where stalk joins; submerged leaves thin, lettuce-like.
Fruit: bottle-shaped, ripening above water.
Uses: Water-lilies, not surprisingly, were held to have cooling properties. Gerard's *Herbal* of 1597 states, 'the root of the Yellow cureth hot diseases of the kidnies and bladder.'

FUMITORY FAMILY Fumariaceae

Fumitories are sprawling weeds of cultivated or disturbed ground. The leaves are finely cut, reminiscent of Maidenhair Fern. They are hairless, often greyish-green and waxy. The small flowers are tubular and 2-lipped.

COMMON FUMITORY
Fumaria officinalis

Common Fumitory is a slender, glaucous, long-stemmed annual, climbing or almost erect. It is a weed of arable land and prefers light soils.
Flower: purplish pink, tips blackish red; tube-like corolla 7-9 mm long, spurred behind, 2-lipped and consisting of 2 outer petals and 2 narrow inner ones; 2 very small sepals.
Flower arrangement: raceme, usually of more than 20 flowers, longer than its stalk.
Flowering time: May-October.
Leaf: all arising from stem, finely divided into flat segments.
Fruit: more or less spherical, 2-3 mm.

CABBAGE FAMILY Cruciferae

The members of this large family, with some 3,500 species, typically have 4-petalled flowers resembling a cross – hence the Latin name – and 6 stamens. The flowers are very often yellow or white and carried in racemes, tightly packed at first but elongating after the first flowers open. The fruits are commonly long and narrow. The family contains many important vegetables, such as turnips and cabbages, as well as many weeds.

GARLIC MUSTARD
Alliaria petiolata

A rather tall, upright biennial, unbranched, up to 120 cm, Garlic Mustard smells faintly of garlic when crushed. The plant can be seen along hedgerows, walls and wood margins.
Flower: white, 6 mm across.
Flowering time: April-June.
Leaf: pale, rather bright green, heart-shaped, toothed; shape reminiscent of nettle leaf.
Fruit: long, very narrow capsule, 2-7 cm long, opening by 2 valves from below.

SHEPHERD'S PURSE
Capsella bursa-pastoris

Hardly a walk can be taken, whether in town or country, without seeing Shepherd's Purse, an annual 3-40 cm high, of wayside, cultivated land and waste places. Distributed throughout the world, having followed man as a weed of agriculture.
Flower: white, about 2·5 mm across. crowded at the top of a raceme, becoming less dense in fruit.
Flowering time: all the year.
Leaf: basal leaves in rosette, deeply lobed or not; base of upper leaves clasping stem.
Fruit: heart-shaped, 6-9 mm across, the 2 valves breaking apart to disperse pale brown seeds – the 'money' in the 'purse'.

CUCKOO FLOWER
Cardamine pratensis

A very pretty plant of damp places and stream sides, the Cuckoo Flower or Lady's Smock, reaches 60 cm.
Flower: lilac, pink or white, 12-18 mm across; petals may be notched.
Flowering time: April-June.
Leaf: pinnate, up to 7 oval or circular leaflets; end leaflet enlarged; basal leaves in rosette.
Fruit: 2·5-4 cm, thin, on long stalks opening from below by 2 valves.

COMMON SCURVY GRASS
Cochlearia officinalis

A common coastal plant the scurvy grass got its name because sailors used to eat it as a source of vitamin C to combat scurvy. With rather lax and often sprawling reddish stems, it reaches 50 cm.
Flower: white, rarely pale lilac, about 1 cm across. Fragrant. May-August.
Leaf: deep shiny green, hairless and rather fleshy: basal leaves in a rosette, more or less heart-shaped on long stalks. Stem leaves oblong or almost triangular, mostly clasping stem.
Fruit: globular, 3-7 mm across.

WOAD
Isatis tinctoria

Famed as the source of the blue dye used by the ancient Britons, the woad is a native of southern and central Europe, where it grows abundantly on waste ground and roadsides. A much-branched biennial, it reaches 1·5 m.
Flower: yellow, up to 4 mm across, in large branched racemes (panicles). June-August.
Leaf: bright green in basal rosette, slightly hairy, with wavy edges; lanceolate, to 15 cm long. Upper leaves arrow-shaped, grey-green and hairless; clasping stem.
Fruit: dark brown and winged, with one seed like a small ash key: hanging in dense clusters.

HEDGE MUSTARD
Sisymbrium officinale

Hedge Mustard, familiar if not pleasing to the eye, grows in waste places in town and country to a height of 90 cm. An annual, it is stiffly erect and branched above, like candelabra.
Flower: pale yellow, small.
Flower arrangement: in dense racemes.
Flowering time: June-July.
Leaf: lower leaves in a rosette, deeply lobed; end lobe larger than rest.
Fruit: very thin, cylindrical, 10-20 mm, straight, pressed to the stem, which elongates in fruit.

WATER-CRESS
Rorippa nasturtium-aquaticum

Water-cress grows in masses in wet places where there is fresh, moving water. It is 10-60 cm high, hairless, creeping, often rooting: ascending at the tip.
Flower: white, 4-6 mm across, petals about twice as long as sepals.
Flower arrangement: clustered at the top of a raceme.
Flowering time: May-October.
Leaf: pinnate, lower leaves with 1-3 circular or broadly ovate leaflets; upper leaf with 5-9.
Fruit: 13-18 mm, curving slightly upward.
Uses: commercial cultivation as a salad crop started at the beginning of the 19th century.

CHARLOCK
Sinapis arvensis

Charlock grows up to 80 cm and often has stiff hairs on the base of the stem. It is a serious weed of arable land. An annual.
Flower: bright yellow, 1·5-2 cm; petals narrowed at base.
Flower arrangement: dense raceme, spreading as fruit ripens.
Flowering time: May-July.
Leaf: up to 20 cm, alternate, roughly hairy; lower leaf stalked, deeply lobed; end lobe large, coarsely toothed; upper leaves stalkless, without lobes.
Fruit: 2·5-4·5 cm long, narrow, abruptly narrowed at tip, indented between seeds, smooth or stiffly hairy.

lower leaf

VIOLET FAMILY Violaceae

The flowers in this family are markedly irregular, with 5 petals of which the lowest has a long spur at the back. The sepals have small flaps. After producing normal flowers in spring, violets often produce flowers which do not open and which fertilize themselves. The fruits are rounded capsules which split open into three parts.

SWEET VIOLET
Viola odorata

Although cultivated in gardens and often escaping, the scented Sweet Violet is also a native of Europe. It grows in hedgebanks, scrub and woods, and is 5-15 cm high.
Flower: deep violet or white with lilac spur, about 1·5 cm, scented; petals 5, unequal, lower petal forming spur behind; spur longer than sepal appendages; sepals rounded at tip.
Flower arrangement: solitary, on long stalks.
Flowering time: February-April.
Leaf: heart-shaped, rounded or pointed at tip, blunt-toothed, sparsely hairy, long-stalked.

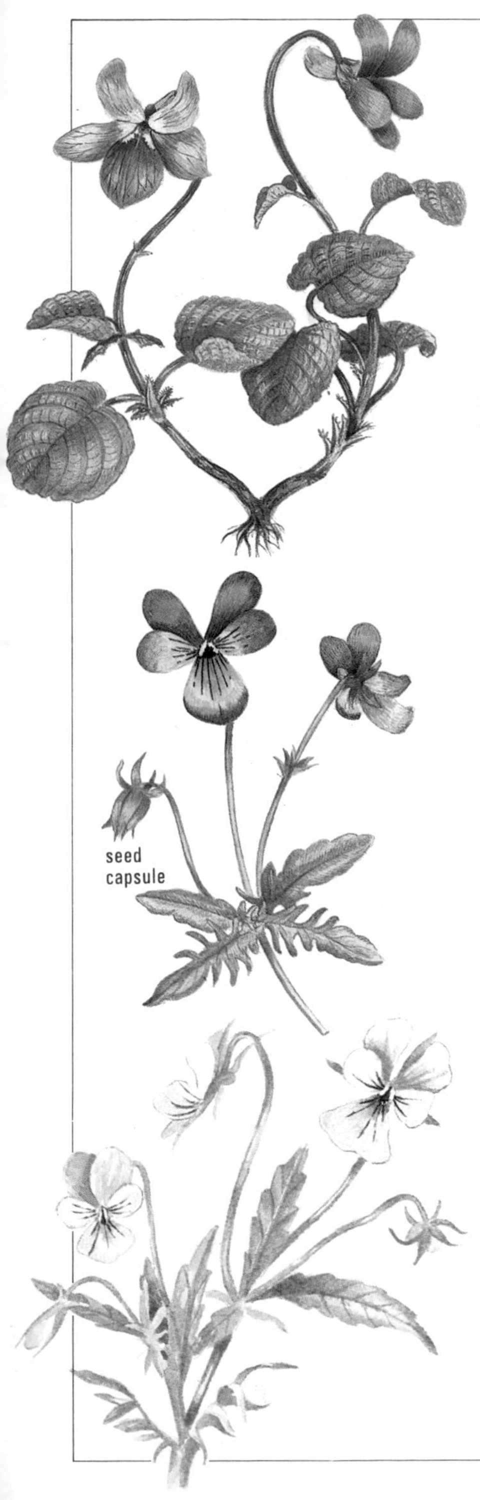

COMMON DOG-VIOLET

Viola riviniana

This violet can be very variable in size. It lacks creeping stems and grows in woods, on hedgebanks or in grassland.
Flower: usually blue-violet though varying; spur paler; lower petal with many long, dark veins; unscented, 1-2·5 cm; spur 2-5 mm, stout, often curved up, notched at tip; sepals with appendages about 2-3 mm.
Flower arrangement: solitary.
Flowering time: April-June.
Leaf: in loose rosette; stalked, heart-shaped, blunt-toothed, up to 8 cm; stipules toothed.

WILD PANSY

Viola tricolor

This ancestor of the garden pansy grows on cultivated and waste ground and in short grassland. It is normally 15-45 cm high, but a dwarf form 3-15 cm high grows on dunes and grassland by the sea. Grassland plants are normally perennial, but those growing on disturbed land are annuals.
Flower: blue-violet or yellow, or a mixture of the two: scentless; 1-3 cm vertically. Petals usually longer than sepals, the latter being pointed and rather shorter than spur.
Flower arrangement: solitary, several arising from each stem.
Flowering time: April-September.
Leaf: alternate; lower leaf ovate or heart-shaped, blunt-toothed; upper leaf narrower; stipules leaf-like deeply lobed.

FIELD PANSY

Viola arvensis

This low-growing annual is a common weed of cultivated land, especially on chalky and neutral soils.
Flower: creamy white, sometimes tinged with yellow or violet. 1-1·5 cm across. Petals usually shorter than sepals.
Flower arrangement: solitary on long stalks springing from axils.
Flowering time: April-November.
Leaf: basal leaves rather broad; stem leaves narrow, with deeply-toothed leafy stipules.

MILKWORT FAMILY Polygalaceae

This world-wide family includes shrubs and small trees but all the North European members are small plants growing in short grassland. The 3 true petals of the flowers are enclosed within the 5 petal-like sepals.

COMMON MILKWORT
Polygala vulgaris

This is a rather low, much branched plant, 10-30 cm high, to be found in short grassland, heaths and dunes. The tiny flowers are almost enclosed by 2 enlarged, coloured sepals.
Flower: blue, pink or white, 6-8 mm; sepals 5, the two inner enlarged, coloured, petal-like.
Flower arrangement: a raceme.
Flowering time: May-September.
Leaf: alternate, narrow, entire, more or less pointed; lower leaves 5-10 mm, upper leaves longer.
Fruit: flat, heart-shaped capsule.
Uses: herbalists formerly prescribed this little plant for nursing mothers, to increase the flow of milk.

ST JOHN'S WORT FAMILY Hypericaceae

A family of approximately 1,000 species. Glands, appearing as black or translucent dots on the plants, are characteristic of this family. There are 5 petals, and the many stamens are joined at their bases into bundles.

PERFORATE ST JOHN'S WORT
Hypericum perforatum

There are tiny translucent dots (seen if the plant is held to the light) and black dots to look for on this plant. Running down the stem are 2 raised lines. The plant is hairless, erect and 10-100 cm high, growing in grassland and scrub.
Flower: yellow, about 2 cm across; petals 5, tiny black dots around margins; sepals 5, pointed, with or without black dots; stamens many.
Flower arrangement: branched cyme.
Flowering time: June-September.
Leaf: alternate, with many translucent dots; 1-3 cm, elliptic to linear.

ROCKROSE FAMILY Cistaceae

This family of about 250 herbs and small shrubs is centred on the Mediterranean region, where the plants thrive on dry, sunny slopes. There are generally 5 petals, numerous stamens, and 5 strongly-veined sepals of which three are much larger than the other two. Individual flowers are short-lived, with the petals falling after just one day.

COMMON ROCKROSE
Helianthemum chamaecistus

This creeping, mat-forming plant has a sturdy woody base from which spring many slender branches – soft at first but becoming wiry. It reaches 30 cm and is best described as an undershrub. It grows on grassy and rocky slopes, especially on limestone, in many parts of Europe.

Flower: bright yellow, 2·5 cm across; petals creased at first. Up to 12 flowers in a loose cyme with downy stalks. May-September.

Leaf: oval, to 2 cm long: deep green above, pale and furry below.

Fruit: an egg-shaped capsule splitting into three lobes.

STITCHWORT FAMILY Caryophyllaceae

This large family includes Carnations, Pinks and many other garden plants, as well as widespread weeds such as the Chickweeds. The flowers have 4 or 5 petals, which are notched, except in the Sandworts and Spurreys. Stamens are normally twice as numerous as petals. Some flowers open or emit scent only at night to attract moths for pollination.

SEA SANDWORT
Honkenya peploides

The Sea Sandwort is a small, fleshy plant, 5-25 cm high, partly creeping on sand and shingle by the sea. It can withstand brief immersion in salt water.

Flower: greenish white, 6-10 mm; petals 5, rounded, as long as or shorter than the sepals.

Flower arrangement: solitary, in leaf axils and forks of branches.

Flowering time: May-August.

Leaf: opposite, ovate, fleshy, stalkless. Bright green.

Fruit: capsule, spherical, opening by 3 teeth; seeds large, pear-shaped, reddish-brown.

CORN SPURREY
Spergula arvensis

Corn Spurrey is a pale green annual, 5-70 cm high, slightly to very sticky, with weak, branched stems and whorls of leaves. It is a weed of cultivated land, preferring sandy soils.
Flower: white, 4-7 mm across; petals 5, rounded, slightly longer than sepals.
Flowering time: June-August.
Leaf: 1-3 cm long, needle-shaped, in whorls, slightly fleshy, channelled beneath, sticky.
Fruit: capsule, 5 mm, opening by 5 teeth; seeds blackish, warty.
Uses: various forms of this weed are grown in Germany and the Netherlands as a nutritious fodder crop for sheep and cows. When cultivated it may grow 90 cm high.

SAND SPURREY
Spergularia rubra

Sand Spurrey, with its tiny, pink flowers, is a straggling plant, 5-25 cm high, found on sandy soils in open ground. The upper parts of the plant have slightly sticky hairs. An annual.
Flower: pink, paler at base of petals, 3-5 mm, petals 5; sepals longer than petals.
Flower arrangement: few-flowered cyme.
Flowering time: May-September.
Leaf: in clusters along stem, very narrow, tapering to stiff, sharp point, not fleshy; stipules silvery, conspicuous.
Fruit: capsule, 4-5 mm, opening by 3 teeth; seeds tiny, dark brown.

WHITE CAMPION
Silene alba

The White Campion is a branched, softly hairy weed of cultivated land and grows to 80 cm. As with the Red Campion, the flowers are unisexual. They open in the evening when they are slightly scented to attract moths as pollinators.
Flower: white, 2·5-3 cm across; petals 5, deeply cleft, narrowed at base; calyx-tube hairy, sticky.
Flower arrangement: cyme.
Flowering time: May-September.
Leaf: opposite, ovate, stalked; stem leaves stalkless.
Fruit: capsule 1-1·5 cm, ovoid, opening by 10 teeth.

seed
capsule

RED CAMPION
Silene dioica

The Red Campion grows in woods and hedgerows. The whole plant is softly hairy and grows up to 80 cm high. There are separate male and female flowers. Red Campion interbreeds readily with White Campion, the offspring having pale pink flowers.
Flower: bright rose, 18-25 mm across; petals 5, deeply cleft, narrowed at base; calyx-tube hairy, slightly sticky.
Flower arrangement: cyme.
Flowering time: May-June.
Leaf: opposite, broadly ovate; the basal leaf blade continuous as thin border down each side of stalk; upper leaves stalkless.
Fruit: capsule 1-1·5 cm, spherical or ovoid, opening by 10 teeth which curve back.
Uses: Campions were associated with snakes, the pounded seed being used to treat snake bite.

BLADDER CAMPION
Silene vulgaris

This erect, often grey-green, branching plant is usually without hairs and grows up to 90 cm high. The 'bladder' is the inflated calyx-tube. It is found in grassland and arable land, and along roadsides.
Flower: white, about 1·5 cm across; petals 5, deeply cleft, narrowed at base; calyx-tube inflated, net-veined; bracts papery.
Flower arrangement: loose cyme.
Flowering time: June-August.
Leaf: often greyish-green, opposite, ovate; lower leaves short-stalked, upper stalkless.
Fruit: capsule, with 6 erect teeth, enclosed by persistent calyx.
The Sea Campion (*S. maritima*), growing on cliffs and shingle, is similar but shorter. Its greyish leaves form dense mats. Petals are broader.

COMMON MOUSE-EAR
Cerastium holosteoides

The flowering stems of this very common little plant reach 45 cm. It is found in grassland and waste places almost everywhere, growing in loose, straggling tufts. The leaves and stem are hairy.
Flower: white; petals 5, deeply notched, equalling sepals; sepals with narrow, papery margins and showing between petals of open flower; stamens 10 or 5.
Flower arrangement: loose clusters.
Flowering time: April-September.
Leaf: 10-25 × 3-15 mm, stalkless, tip pointed or rounded.
Fruit: capsule, curved; seeds warty.

lower leaf

RAGGED ROBIN
Lychnis flos-cuculi

The rose-red, finely dissected flower petals of the Ragged Robin are striking. The stem and leaves are slightly rough to the touch. The height of the branched stem is 20-90 cm. The plant is a lover of damp places.
Flower: rose-red, 3-4 cm across; petals deeply cut into thin segments; calyx-tube 5-toothed; stamens 10.
Flower arrangement: cyme.
Flowering time: May-June.
Leaf: opposite; basal leaves ovate, stalked, slightly rough to touch; upper leaves stalkless.
Fruit: capsule 6-10 mm, opening by 5 teeth.

GREATER STITCHWORT
Stellaria holostea

The slender, flowering stems are 15-60 cm high and 4-angled, the angles being slightly rough to the touch. The plant is slightly greyish-green and grows in hedgerows and woods.
Flower: white, 2-3 cm, long-stalked; petals 5, notched to about half their length; sepals with narrow, papery margins; stamens 10.
Flower arrangement: loose forking cymes.
Flowering time: April-June.
Leaf: rather greyish-green, opposite, 4-8 cm long, lanceolate, tapering to sharp point, stalkless.
Fruit: capsule, spherical; seeds reddish-brown, warty.
Uses: seed of Stitchwort was powdered, added to wine and used against pains or 'stitches' in the side.

COMMON CHICKWEED
Stellaria media

The straggling stems of this world-wide annual weed are 5-40 cm long. The tiny flowers are inconspicuous and there is a single line of hairs running down the stem.
Flower: white, 8-10 mm; petals very deeply cleft; sepals same length as petals with narrow, papery margins, showing between petals of open flower.
Flower arrangement: loose clusters.
Flowering time: all year.
Leaf: opposite, ovate; lower leaves stalked, upper stalkless.
Fruit: capsule, stalk curved down.
Uses: young plants can be eaten in salads and sandwiches. Small birds love the seed.

ROCK SOAPWORT
Saponaria ocymoides

This sprawling, hairy plant grows in the Alps and Pyrenees, where it clothes many rocky and stony slopes with beautiful pink cushions or carpets throughout the summer. It can be found at altitudes up to 2,400 m.
Flower: deep pink to purple 6-10 mm across. Calyx tube long and reddish-brown and clothed with glandular hairs. In dense clusters. March-October.
Leaf: greyish green and hairy; oval or spoon-shaped up to 1 cm long.

MESEMBRYANTHEMUM FAMILY Aïzoaceae

The 1,200 or so members of this family come from the warmer parts of the world, especially from South Africa. They are herbs and undershrubs and all are adapted for life in dry places. Many have fleshy leaves. The flowers often look like large daisies, but they are not related: each is a single flower with many petals, and not a composite head (see pages 20-1). One of the best known members of the family is the Mesembryanthemum itself, whose brilliant flowers make it a favourite garden plant.

HOTTENTOT FIG
Carpobrotus edulis

Introduced from South Africa, this striking plant has become naturalized on many parts of the European coastline. It is a sprawling plant, 25-30 cm high, forming dense carpets on cliffs and shingle banks.

Flower: deep pink, mauve or yellow, up to 5cm across, with many strap-shaped petals and many stamens; calyx tube with 5 leafy lobes. Borne singly on swollen stalks. April-July.

Leaf: bright green, fleshy and sausage-shaped; to 10 cm long and triangular in cross-section.

Fruit: a fleshy capsule: edible. Several closely related species have now been introduced to various parts of the European coastline.

GOOSEFOOT FAMILY Chenopodiaceae

This family contains some 1,500 species, including such economically important plants as beetroot and spinach. The flowers are inconspicuous, clustered into slender spikes and wind-pollinated. The leaves of many species have a pale mealy coating composed of minute bladder-like hairs. Many are salt-loving species, growing in both moist coastal habitats and desert regions. Common saltmarsh and seashore species include Glasswort, Sea Beet, and Sea Purslane. There are also many common agricultural weeds in the family, most of them annuals.

FAT HEN
Chenopodium album

Growing 10-150 cm high, Fat Hen often has a red tinge to the grooved stem. The plant is deep green with a mealy covering and grows on cultivated and waste land. It is very variable and there are several other similar species. An annual.
Flower: green, inconspicuous, bisexual.
Flower arrangement: spikes, arising from leaf-axils.
Flowering time: July-October.
Leaf: alternate, mealy, diamond-shaped to narrowly ovate, stalked, toothed, sometimes nearly 3-lobed or entire.
Fruit: black achene.

flower

lower leaf

COMMON ORACHE
Atriplex patula

The plant is branched, up to 150 cm high and slightly mealy. The stems are ridged and striped white and green or red and green. This is a weed of waste places and cultivated ground. Although similar to Fat Hen, Common Orache can be distinguished by its separate male and female flowers. An annual.
Flower: green, inconspicuous, unisexual, males with 5 green 'petals' and 5 stamens, females with 2 stigmas and enclosed by 2 diamond-shaped bracts.
Flower arrangement: spikes arising from leaf-axils.
Flowering time: July-September.
Leaf: alternate, diamond-shaped, base tapering into leaf-stalk, toothed or entire; upper leaves narrow, entire, stalkless.
Fruit: black, enclosed by 2 bracts.

lower leaf

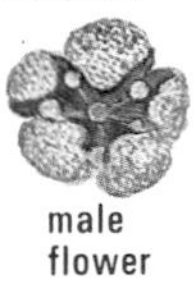

MALLOW FAMILY Malvaceae

Mallows, Cotton and Hollyhock all belong to this family. The 5 petals are twisted in the pointed bud, and the many stamens are bunched on a central tubular stalk. There is an epicalyx, usually of three segments, below the 5 true sepals. The plants are downy and a lens reveals that the hairs of most species are branched and star-shaped.

COMMON MALLOW
Malva sylvestris

This very variable plant is erect or straggling up to 150 cm high, and the stem is woody at the base. The plant favours dry, open places on waste and cultivated land.
Flower: rose-purple, dark-striped, 2·5-4 cm across; petals 5, thin, narrowed at base, notched; several stalked flowers arising from each leaf-axil.
Flowering time: June-September.
Leaf: palmately lobed, blunt-toothed, often folded along main veins; basal leaves long-stalked.
Fruit: nutlets, arranged in circle, covered by network of ridges.

GERANIUM FAMILY Geraniaceae

Many species in the Geranium family have long, soft hairs, and the flowers are usually brightly coloured, with 5 petals. The characteristic fruit is arranged around the base of the stiff, persistent style and splits into 5 one-seeded sections when ripe.

MEADOW CRANESBILL
Geranium pratense

This handsome and rather hairy plant grows in grassy places and reaches 100 cm (sometimes more in damp places). It occurs in many parts of Europe, especially on limestone.
Flower: violet-blue, or occasionally sky-blue, with paler veins; saucer-shaped 25-35 mm across. Paired in loose cymes. June-September.
Leaf: lower ones up to 15 cm across, palmately lobed, each of the 5-7 lobes deeply toothed: upper leaves smaller and 3-lobed.
Fruit: hairy, drooping at first and becoming upright as it ripens.

HERB ROBERT
Geranium robertianum

The stem and leaves of Herb Robert are hairy and often turn bright red. The plant grows 10-50 cm high, is branched and erect or straggling. If bruised, it smells unpleasant. It is most often found in hedgerows but also grows among rocks, on walls and in woodland.
Flower: bright pink-mauve, 2 cm across, long-stalked; petals 5, narrowed at base; sepals erect, with bristle-like tips.
Flowering time: May-September.
Leaf: in 3 main segments, each deeply and repeatedly dissected; long-stalked.
Fruit: arranged around base of stiff, persistent style to which segments are attached by thin strands.

COMMON STORK'S-BILL
Erodium cicutarium

This plant looks rather similar to Herb Robert but differs in the clustered flowers, the pinnate leaf-segments and the spiralling strands attached to the fruits. The stem and leaves are slightly to very hairy. The plant grows up to 60 cm high in dry, open places, particularly near the sea.
Flower: bright rose-purple; petals 5, 12-14 mm across, upper 2 sometimes with black spot at base, sepals spreading, with bristle-like tips. 5 of the 10 stamens lack anthers.
Flower arrangement: 3-12 flowers in umbel-like cluster.
Flowering time: June-September.
Leaf: doubly pinnate.
Fruit: arranged around base of stiff, persistent style to which segments attached by thin, spiralling strands.

WOOD SORREL FAMILY Oxalidaceae

The flowers in this family have 5 petals and 10 stamens. The family is mainly distributed in the tropics. Some *Oxalis* species are grown as ornamentals in gardens, and a few have edible tubers.

WOOD SORREL
Oxalis acetosella

A small plant of woodland and shaded hedgebanks, Wood Sorrel has leaves composed of 3 leaflets, like clover leaves but a paler, more delicate green. In cold weather and at night the leaves fold down.
Flower: white, lilac-veined, bell-shaped, nodding; petals 5, 8-15 mm long; sepals 5.
Flower arrangement: solitary.
Flowering time: April-May.
Leaf: in 3 rounded, notched leaflets, pale yellow-green.
Fruit: capsule, rounded, 5-angled, 3-4 mm long; seeds light brown.

PEA FAMILY Leguminosae

The flowers in this very large family are easily recognizable, being composed of an upright 'standard' petal, 2 side 'wing' petals and a pair of fused, boat-shaped petals under the standard, as in the Sweet Pea. The 5 sepals are fused into a calyx tube, and stamens and style are usually enclosed by the petals. The flowers are loosely clustered or in small, dense heads. Leaves are usually trifoliate or pinnate, often with a tendril at the end. The fruit is typically a pod, and many important food crops like soybeans and peas belong to this family. Clover and Lucerne are very widely grown as fodder and green manure. The roots form nodules which contain nitrogen-fixing bacteria and act as a natural fertilizer.

COMMON REST-HARROW
Ononis repens

A shrubby, hairy plant of 40-70 cm, this species is prostrate, often with upturned tips bearing the pink flowers. The stems are hairy all round. The plant grows in rough grassland and dunes.
Flowers: pink, 1·5-2 cm long, on shortly hairy stalks; wing petals as long as keel; calyx tube with 5 long teeth, densely hairy.
Flowering time: June-September.
Leaf: 1-3 leaflets, oval, finely toothed, hairy, short-stalked.
Fruit: pod, hidden by calyx.

COMMON BIRD'S-FOOT TREFOIL
Lotus corniculatus

This plant is usually almost hairless and grows in grassland. Plants growing on the coast may have small, fleshy leaves. The prostrate stems, 5-35 cm long, turn up at the tips.
Flowers: yellow, often tinged or streaked with red, 1-1·5 cm long.
Flower arrangement: 2-7 flowers in outward-facing ring on erect stalk
Flowering time: June-September.
Leaf: of 5 elliptic or almost circular leaflets, each 3-10 mm long, rounded at tip or with a short point; short leaf-stalk often between lower and upper leaflets.
Fruit: pod, 1·5-3 cm long, cylindrical, straight; spreading out from stem like the toes of a bird – hence the common name.

RIBBED MELILOT
Melilotus officinalis

A native of southern Europe, but now widely distributed, this rather slender, hairless biennial is often abundant on roadsides and other bare or waste places: also on dunes and as a weed of cultivation. It reaches 120 cm.
Flower: bright yellow, 5 mm long: in slender, tapering racemes up to 50 mm long. May-September.
Leaf: trifoliate, each leaflet being oval, up to 20 mm long, and strongly toothed.
Fruit: a brown oval pod to 5 mm long.

SAINFOIN
Onobrychis viciifolia

A rather downy and generally upright herb, the Sainfoin resembles a small lupin when in flower. It reaches 50 cm and occurs widely in dry grassland and on roadsides in southern half of Europe: widely cultivated for fodder.
Flower: 10-12 mm across; deep pink with darker veins. In dense conical spikes. May-September.
Leaf: pinnate, with 6-12 pairs of leaflets each up to 3 cm long.
Fruit: a semi-circular, one-seeded pod that does not split open.

BLACK MEDICK
Medicago lupulina

The leaves of Black Medick are clover-like and the flowers yellow.
The plant is sprawling, 5-50 cm high and downy, growing in open places such as roadsides, cultivated fields and at the coast. It is an annual or short-lived perennial.
Flower: bright yellow, 2-3 mm.
Flower arrangement: grouped into small, round heads 3-8 mm across, on stalks longer than the leaf-stalks.
Flowering time: April-August.
Leaf: of 3 leaflets, each narrowed towards base, finely toothed, often shallowly notched at tip in which is a minute bristle; stipules finely toothed.
Fruit: tightly curved disc, about 2 mm across, containing 1 seed, covered by a network of ridges. Black when ripe and clustered tightly at top of flower stalk.
Uses: this Medick is sometimes grown for fodder.
The Lesser Trefoil (*Trifolium dubium*) is very similar to Black Medick, but is an almost hairless annual. Its fruits are brown and straight and always covered with the dead petals.

HORSESHOE VETCH
Hippocrepis comosa

Spreading out from a woody rootstock, this plant forms beautiful carpets of yellow when in flower on chalk and limestone grasslands and on limestone rocks. It reaches heights of about 20 cm. It is much like the Bird's-foot Trefoil (page 49) but its leaves are longer and the characteristic seed pods easily distinguish it.
Flower: bright yellow, 6-10 mm long: up to 10 in a flat-topped whorl at top of stem up to 10 cm long. May-July.
Leaf: pinnate, to 8 cm long: 4 or 5 pairs of leaflets plus a terminal one, all oblong and blunt-ended (not pointed as in Bird's-foot Trefoil).
Fruit: a cluster of pods radiating from top of stem: each pod to 30 mm long and strongly waved; breaking up into several horseshoe-shaped, one-seeded sections.

KIDNEY VETCH
Anthyllis vulneraria

Clothed with silky hairs, this is a rather variable species – sometimes prostrate and creeping, at other times erect and reaching 30 cm. Normally a perennial, but occasionally behaving as an annual, it grows on calcareous grasslands, dunes, and cliffs.
Flower: normally yellow, sometimes orange; occasionally red, pink, or white by the sea. Tightly packed into rounded or kidney-shaped heads 2-4 cm across. June-September.
Leaf: pinnate, to 6 cm long, very silky below. 4 or 5 pairs of linear leaflets plus a terminal one, the latter large and lanceolate in lower leaves and linear in upper leaves.
Fruit: a flat pod about 3 mm long.

COMMON VETCH
Vicia sativa

The Common Vetch is a slightly hairy plant, scrambling or climbing in hedges or grassy places. It can grow up to 120 cm.
Flower: purplish red, 1-3 cm long; calyx teeth about equal.
Flower arrangement: solitary or in pairs, on very short stalks or stalkless.
Flowering time: May-September.
Leaf: pinnate, with 3-8 pairs of narrow leaflets, each 1-2 cm long, often wider at tip than base, with minute bristle, notched or pointed; tendrils branched or not; stipules toothed or not, often with dark spot.
Fruit: pod, 2·5-8 cm long, yellow-brown to black, hairless or slightly hairy; seeds 4-12.
Uses: a form of this Vetch is often grown as a fodder crop.

TUFTED VETCH
Vicia cracca

Tufted Vetch is a weak-stemmed plant which gains support by clambering over other vegetation. It is 60-200 cm long, slightly hairy and grows in hedges and bushy places. The long, crowded racemes of flowers are characteristic of this species.

Flower: blue-purple, 8-12 mm long, each drooping on short stalk; style equally hairy all round; upper teeth of calyx-tube minute; lower teeth about as long as calyx-tube.

Flower arrangement: dense raceme of 10-40 flowers on stalk 2-10 cm long.

Flowering time: June-August.

Leaf: pinnate, with 6-15 pairs of narrow leaflets, each 1-2·5 cm long, pointed or with minute bristle; leaf ends in branched tendril; stipules untoothed.

Fruit: pod, brown, hairless, 1-2·5 cm long; seeds 2-6.

BUSH VETCH
Vicia sepium

Climbing or scrambling in grassy or bushy places, the Bush Vetch grows 30-100 cm long. The leaves and stem have short hairs or are almost hairless.

Flower: pale purple, 1-1·5 cm; lower calyx teeth much shorter than calyx-tube, and tips curving toward each other.

Flower arrangement: raceme of 2-6 flowers, very shortly stalked.

Flowering time: May-August.

Leaf: pinnate, with 3-9 pairs of leaflets, each 1-3 cm long, variably shaped; tips rounded or pointed, with or without minute bristle at centre of slight notch at leaflet-tip; tendrils branched; stipules sometimes toothed, with black spot.

Fruit: pod, 2-3·5 cm long, black, hairless; seeds 3-10.

MEADOW VETCHLING
Lathyrus pratensis

This Vetchling, with or without short hairs, scrambles over vegetation in grassy or bushy places. It is 30-120 cm in length and has a sharply angled stem.
Flower: yellow, 10-18 mm long; teeth of calyx-tube narrowly triangular.
Flower arrangement: raceme of 5-12 flowers on long stalk.
Flowering time: May-August.
Leaf: 1-2 pairs of narrow, pointed leaflets, 1-3 cm long; tendril branched or not; stipules leaf-like, 1-2·5 cm, arrow-shaped.
Fruit: pod, 2-4 cm long, black; seeds 5-12.

WHITE CLOVER
Trifolium repens

A very familiar little plant of grassland, including lawns and verges. The stems of White Clover creep along the ground rooting at intervals. The plant is hairless and may grow as long as 50 cm.
Flower: white or pink-tinged, scented, individual flowers 8-13 mm, turning down and becoming brown from base of head upwards; calyx-tube white, green-veined with narrow, pointed teeth.
Flower arrangement: round head on long, grooved stalk.
Flowering time: June-September.
Leaf: of 3 rounded leaflets, minutely toothed, each 1-3 cm long, each often with pale crescent; stipules joined, sheathing stem, with 2 free points.
Fruit: pod, hidden by dead flower.
Uses: bees use the nectar to produce clover honey, and several varieties are grown for fodder.

RED CLOVER
Trifolium pratense

Red Clover grows erectly 5-100 cm high in grassland, and is more or less hairy.
Flower: pink-purple; individual flowers 15-18 mm long.
Flower arrangement: in compact, rounded or ovoid heads of 2-4 cm.
Flowering time: May-September.
Leaf: of 3 elliptic leaflets, each 1-3 cm long, often with pale crescent, usually rounded at tips; leaf-stalk up to 20 cm; pair of short-stalked leaves immediately below flower head; stipules conspicuous, joined up the leaf-stalk, with 2 free points, each ending in bristle.
Fruit: pod, about 2 mm.
Uses: several varieties of red clover are grown for hay and silage, usually in mixed stands with various grasses. They are also sown with grasses to produce a good sward for new road-side verges, the nitrogen-fixing abilities of the clovers helping to nourish the grasses.
Several similar species of clover can be seen growing wild and in meadows. Zig-zag Clover (*T. medium*) has deeper red flowers and much brighter leaves than Red Clover, with at most a small white spot on each leaflet, and its stem is angled at each node to produce a zig-zag effect. Long-headed Clover (*T. incarnatum*) has very hairy, cylindrical flower heads up to 20 mm long, with the pale pink or cream flowers almost hidden among the hairs. It grows mainly on sea cliffs. A widely cultivated variety of this species, known as Crimson Clover, usually has bright crimson flowers on elongated heads up to 40 mm long.

BALSAM FAMILY Balsaminaceae

The Balsams are hairless plants with juicy, easily broken, translucent stems often with stalked, sticky glands. The flowers are irregular in shape, with 5 petals forming a broad lip, a hood and a curved spur behind. Typically the flower dangles from the flower stalk. The fruits explode on being touched.
The yellow flowered Touch-me-not Balsam (*Impatiens noli-tangere*) gets its name from this explosive habit.

INDIAN BALSAM
Impatiens glandulifera

Introduced as a garden plant from the Himalayas in 1839, and naturalized along waterways and in waste places, this tall, stout-stemmed species grows 100-200 cm high. It is hairless and the stems reddish. There is no mistaking the rather orchid-like dangling mauve flowers.
Flower: purplish pink, 2·5-4 cm, petals 5, forming a broad, lower lip and hood; sepals 3, lower forming a mauve, spurred bag. The plant is also known as the Policeman's Helmet because of the flower shape.
Flower arrangement: long-stalked racemes arising from leaf-axils.
Flowering time: July-October.
Leaf: opposite or in threes, 5-18 cm long, elliptic, toothed; reddish glands along basal margins.
Fruit: capsule, club-shaped, opening by 5 valves which spring into coils, shooting out seeds.

ROSE FAMILY Rosaceae

This ancient family is among the more primitive of the flowering plant families. The flowers are often conspicuously coloured, but very rarely blue. The regular, unspecialized flowers attract a wide range of insects. There are normally 5 petals and many stamens. The fruits are varied and include achenes, follicles, drupes (stone-fruit), and the apples and pears, which are false fruits known as pomes.
The popular modern roses of gardens are complicated hybrids, far from their wild ancestors.

WOOD AVENS or HERB BENNET
Geum urbanum

Growing erect in shady places, the Wood Avens is 20-60 cm tall and shortly hairy.
Flower: yellow, 10-18 mm across; petals 5, oval; sepals 10, 5 as long as petals, alternating with 5 shorter; stamens many; styles many, in central tuft.
Flowering time: June-August.
Leaf: basal leaves pinnate, stalked, with 2-3 pairs of toothed leaflets, 5-10 mm long; end leaflet enlarged; stipules leaf-like.
Fruit: like that of Water Avens.

WATER AVENS
Geum rivale

A plant of shady, wet places and cleared woodland, the Water Avens is 20-60 cm high and shortly hairy. The flowers may easily be overlooked due to their nodding habit.
Flower: orange-pink, nodding; petals 5, 1-1·5 cm long, shallowly notched, abruptly narrowed at base; sepals purple, 5 as long as petals alternating with 5 shorter; stamens many; styles many, in central tuft.
Flower arrangement: few-flowered cyme.
Flowering time: May-September.
Leaf: pinnate, with 3-6 pairs of toothed leaflets; end leaflets enlarged, more or less rounded; leaf-stalk long; upper leaves smaller with fewer leaflets.
Fruit: head of achenes; styles persistent, becoming hooked; dispersed by clinging to fur of passing animals.

LADY'S MANTLE
Alchemilla vulgaris agg.

A group of closely related plants, all fairly low-growing and mostly rather hairy perennials of damp meadows and other damp, grassy places, especially in the hills.
Flower: 3-5mm across: greenish yellow with four sepals and no petals.
Flower arrangement: dense clusters of compound cymes.
Flowering time: May-September.
Leaf: palmately lobed, but not cut deeply: green on both surfaces.

CLOUDBERRY
Rubus chamaemorus

A low-growing perennial with rather downy stems and no prickles. Inhabits upland bogs and moors, especially in the north.
Flower: white and very conspicuous, although not flowering freely in Britain. Up to 2·5cm across.
Flower arrangement: erect and solitary; male and female flowers on separate plants.
Flowering time: June-August.
Fruit: red, becoming orange when fully ripe: with fewer drupelets than blackberry.
Leaf: palmate and hairy, with 5-7 lobes especially well developed in male plants.

BRAMBLE
Rubus fruticosus

The stems of these woody shrubs often arch over, rooting at the tip and bearing hooked prickles, some of which may be straight. The Brambles are extremely variable. About 2,000 micro-species may be recognized in the *Rubus fruticosus* group. This huge number is thought to have arisen by a combination of hybridization and self-fertilization.

Flower: pink or white, 2-3 cm across; petals 5; stamens many.
Flower arrangement: solitary or clustered.
Flowering time: May-September.
Leaf: of 3-5 toothed leaflets.
Fruit: the 'blackberry'.
Uses: blackberries have been eaten and enjoyed for thousands of years. Healing powers were associated with the leaves and stems, mainly for relieving burns and swellings.

DOG ROSE
Rosa canina

The Dog Roses are extremely variable shrubs, 1-3 m high, with stems that arch and scramble in hedgerow and scrub. The prickles are hooked.
Flower: pink or white, 4-5 cm across; petals 5, notched; sepals 5, lobed, falling before fruit ripens; stamens many; stigmas in a head.
Flower arrangement: cluster of 1-4 flowers, stalks hairless.
Flowering time: June-July.
Leaf: pinnate, with 2-3 pairs of elliptic, toothed, usually hairless leaflets; stipules joined to leaf-stalk.
Fruit: rose hip, scarlet, about 1·5-2 cm long, spherical or ovoid.
Uses: jelly and syrup may be made from the fruit, rich in vitamin C. More than 500 tons of hips were collected in Britain in 1943 to provide children with rose-hip syrup.

MEADOWSWEET
Filipendula ulmaria

This is a rather tall, erect plant, 60-200 cm high, of wet places, including damp meadows. The small, cream-coloured flowers with long stamens are in frothy, irregular masses. The stems are often reddish.
Flower: cream, scented, 4-10 mm across; stamens many, twice length of petals.
Flower arrangement: in loose, irregular masses.
Flowering time: June-September.
Leaf: pinnate, with 2-5 pairs of large, ovate, toothed leaflets with small leaflets between; usually white and downy beneath.
Fruit: carpels becoming twisted together to form achenes about 2 mm.
Uses: Meadowsweet was formerly used against malaria when forms of that disease were common in undrained lowland areas. Strewn on floors, it also gave out a tangy odour like today's air fresheners.

TORMENTIL
Potentilla erecta

Erect or creeping but not rooting, Tormentil is 10-30 cm high. It grows in grassland and may be found on mountains to a height of about 1,000 m. Not on lime.
Flower: yellow, 7-11 mm across; petals 4, shallowly notched, longer than sepals; sepals 8; stamens many.
Flower arrangement: on long slender stalks in loose cymes.
Flowering time: June-September.
Leaf: of 3 toothed segments; stalked basal leaves in rosette, often withering before flowering; stem leaves stalkless with large, toothed, leaf-like stipules (leaf appearing 5-segmented).
Fruit: achenes, hairless, in a head.

COMMON AGRIMONY
Agrimonia eupatoria

This is an erect plant, 30-60 cm high, with a slender spike of yellow flowers above and densely leafy below. The reddish stems have long hairs, and the plant favours dry roadsides and field edges.
Flower: yellow, 5-8 mm across; petals 5; sepals 5, hairy; ring of hooked bristles immediately below; stamens many.
Flower arrangement: slender spike.
Flowering time: June-August.
Leaf: pinnate, with 3-6 pairs of toothed, elliptic leaflets, 2-6 cm long; hairy, greyish beneath, not glandular; smaller leaflets between; stipules leaf-like.
Fruit: grooved; top covered with hooked bristles which point forward; dispersed by catching onto clothing, or fur of passing animals.

WILD STRAWBERRY
Fragaria vesca

This plant grows 5-30 cm high, with long runners which root and form new plants. It grows in woods and on hedgebanks.

Flower: white, 12-18mm across; petals 5, rounded, not notched, edges almost touching and sometimes overlapping; sepals 10, in 2 layers; stamens many.
Flower arrangement: long-stalked cyme; stalk hairs lie flat.
Flowering time: April-July.
Leaf: divided into 3 toothed segments, each 1-6cm long, ovate, bright green, paler with silky hairs beneath; leaf-stalks long, hairy.
Fruit: strawberry; smaller than cultivated variety but very good to eat.

BARREN STRAWBERRY
Potentilla sterilis

This plant is like the Wild Strawberry but has only short runners and duller, bluish-green leaves. It is easily distinguished in flower by its notched or heart-shaped petals which are widely separated. The fruit does not become fleshy.

SALAD BURNET
Poterium sanguisorba

Abundant on dry grassland, especially on chalk and limestone, this plant has a strong smell of cucumber when crushed and makes a pleasant addition to salads. It reaches 50 cm.
Flower: minute and without petals: 4 greenish sepals. Tightly packed into globular heads about 10 mm across. Upper flowers are female, with purplish-red stigmas; lower ones are male with hanging yellow stamens. Wind-pollinated. May-August.
Leaf: almost hairless; bluish-green. Pinnate with 4-12 pairs of leaflets each up to 20 mm long and strongly toothed. Basal ones form rosette.
Fruit: a cluster of achenes, each with a corky covering.
The Greater Burnet (*Sanguisorba officinalis*) is larger, with elongated maroon flower heads. It grows in damp grassland.

SILVERWEED
Potentilla anserina

The name describes the silvery, silky hairs with which this plant is covered. The leaves form a rosette from which creep rooting runners up to 80 cm long. The plant grows in damp roadside places and meadows.

Flower: yellow, 1·5-2·0 cm across; petals 5, rounded; sepals 10, in 2 layers, outer layer toothed and half as long as petals; stamens many.
Flower arrangement: solitary, on long stalk.
Flowering time: June-August.
Leaf: pinnate, 5-25 cm, with 7-12 pairs of deeply toothed leaflets, alternating with much smaller leaflets; dense, silver, silky hairs beneath or silky both sides.
Fruit: hairless achenes in a head.
Uses: the fleshy roots were eaten in the poorer parts of Britain as a vegetable or made into meal for bread and porridge.

SAXIFRAGE FAMILY Saxifragaceae

Many species of Saxifrage are short, mountain plants and several are commonly grown in gardens. The flowers have 5 petals and sepals and 10 stamens. The leaves often form rosettes.

RUE-LEAVED SAXIFRAGE
Saxifraga tridactylites

This is a slender, often red-tinged plant of 3-8 cm, though sometimes taller, often branched and covered in sticky hairs. In sunshine the ring-shaped nectaries at the flower centres exude glistening nectar. Mainly on dry walls in lowland areas.

Flower: white, 4-6 mm across.
Flower arrangement: loose cyme.
Flowering time: April-June.
Leaf: 3-5-lobed; basal leaves not forming rosette, often withered at flowering time.
Fruit: 2-lobed capsule; seeds brown.

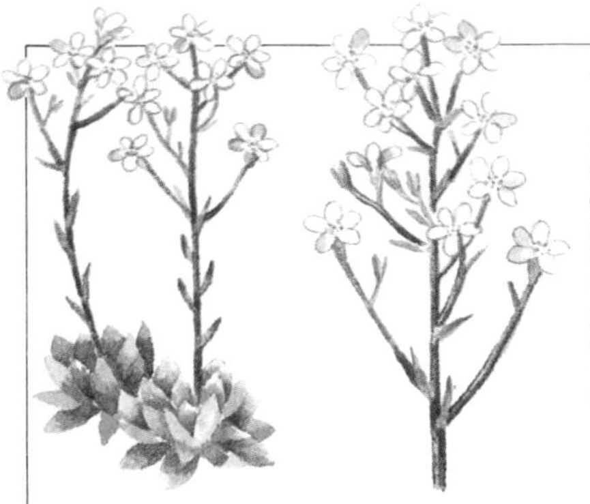

LIVELONG SAXIFRAGE
Saxifraga paniculata

One of several similar mountain-dwellers, this saxifrage occurs on rocks and screes to a height of 2,700 m. Not in Britain.
Flower: white or cream, occasionally pale pink: 8-11 mm across in loose upright or drooping panicles.
Leaf: oblong-oval and encrusted with lime: in open rosettes.

STONECROP FAMILY Crassulaceae

Members of this family are adapted to mountain slopes and other dry habitats such as walls and stony places. Many grow on rocky banks and some grow on sand dunes. They have fleshy (succulent) stems and leaves, all with waxy coatings, and the leaves are usually tightly packed, often forming dense rosettes – all features which help to conserve water. The flowers are usually star-shaped with pointed petals and they are often grouped into broad, domed heads. The fruits are dry and many-seeded.

The two main groups within the family are the stonecrops themselves (*Sedum* spp) and the houseleeks (*Sempervivum* spp). Stonecrops normally have 5 petals, while the houseleeks have 8–16 petals. Several species are grown in rock gardens, where the Ice-Plant (*Sedum spectabile*), with its greyish-green leaves and broad heads of nectar-rich pink flowers, is famed as a butterfly attractant. The Common Houseleek (*Sempervivum tectorum*) was once planted on cottage roofs, where its spreading root system and tightly-packed rosettes helped anchor the slates.

COBWEB HOUSELEEK
Sempervivum arachnoideum

One of the few hairy plants in this family, the Cobweb Houseleek grows on acidic rocks in the Alps and the Apennines, reaching altitudes of over 3,000 m. It is a favourite windowsill plant, for it thrives with little soil. Each rosette produces numerous short, radiating runners which quickly form new rosettes around the parent and soon cover large areas.
Flower: reddish-pink, star-like about 15 mm across; on short stems from the older rosettes. July-September.
Leaf: greyish green, in tight rosettes to 20 mm across: wispy hairs clothe the rosettes and help to conserve moisture. Several closely related, but hairless species of houseleek grow in the mountains of Europe. Most have pink flowers, but there are some yellow-flowered species.

BITING STONECROP
Sedum acre

Growing in mats on dry, stony or sandy ground or on walls, the Biting Stonecrop gets its name from the burning taste of its small, fleshy leaves.
Flower: bright yellow, star-shaped, about 12 mm across; petals 5, pointed; sepals 5; stamens 10.
Flower arrangement: in small clusters at the ends of the branches.
Flowering time: June-July.
Leaf: very fleshy, hairless, rounded, 3-6 mm long, overlapping on non-flowering shoots.
Fruit: group of spreading follicles, each with many tiny seeds.

LOOSESTRIFE FAMILY Lythraceae

This is a small family of plants which favour damp places. The flowers have 4-6 petals which are crumpled in bud. Some species in this family produce dyes, one of the best known being henna.

PURPLE LOOSESTRIFE
Lythrum salicaria

This is an erect plant, 50-150 cm tall with a 4-angled stem and varying from almost hairless to densely hairy. It is found in damp places such as river banks. The flowers are insect-pollinated and are of 3 kinds, differing in the relative lengths of their styles and stamens. Each plant bears only one kind of flower. When bees visit the flowers they pick up pollen from long stamens in just the right position to transfer it to a long stigma of another flower. Similarly, pollen from short or medium length stamens can be transferred to short or medium length stigmas.
Flower: red-purple, 1-1·5 cm across; petals 6, flimsy; calyx tubular, ribbed, with 12 teeth; stamens 12.
Flower arrangement: in whorls on a tall spike.
Flowering time: June-August.
Leaf: opposite or in threes; narrow, pointed, 4-7 cm long, stalkless.
Fruit: capsule, ovoid, enclosed by calyx.

the three style and stamen arrangements

SUNDEW FAMILY Droseraceae

The members of this family are insectivorous plants, trapping and digesting insects with the aid of special leaves. They live in boggy or waterlogged places and the insects supplement the meagre supplies of minerals in such places. The Venus' Fly Trap of America is one of the best known species. Several species of sundew live in Europe.

ROUND-LEAVED SUNDEW
Drosera rotundifolia

Stalked, sticky red glands on the leaves of this little plant often lend a red tinge to the ground of bogs and wet moors where it grows. Insects stick to the glands which slowly bend toward the leaf centre, further enmeshing the insect. Eventually the prey is digested and the glands return to their former positions.
Flower: white, 5 mm across; petals 5 or 6. On a slender spike.
Flowering time: June-August.
Leaf: circular, long-stalked, in a rosette, upper surface covered in sticky, red, stalked glands.

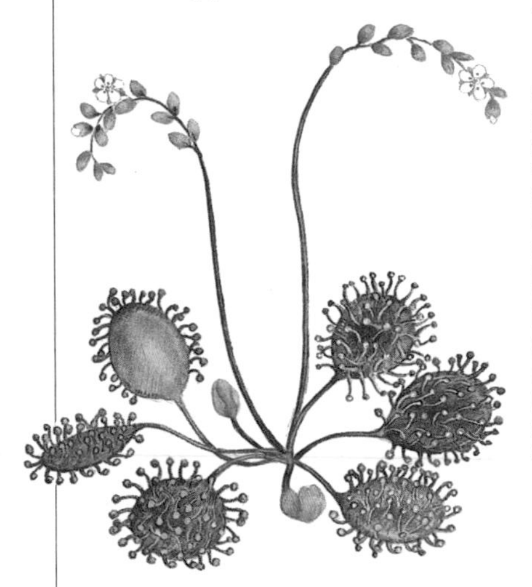

WILLOWHERB FAMILY Onagraceae

This widespread family includes some shrubs and water plants as well as the familiar willowherbs which spring up on roadsides and waste ground. There are also a number of garden plants in the family, including the Evening Primroses (*Oenothera* species). The flowers have 4 petals, which are formed into a tube in the fuchsias. There are commonly 8 stamens and the stigma is often 4-lobed. The fruit is usually a long, slender capsule, which opens by 4 longitudinal slits to release large numbers of plumed wind-scattered seeds.

BROAD-LEAVED WILLOWHERB
Epilobium montanum

This is a small, sparsely hairy plant. The stems are erect, 20-60 cm high and reddish. It is found in gardens, woods, hedges and stony places.
Flower: pale mauve, 6-9 mm across; petals 4, notched; sepals 4, erect, pointed; stamens 8.
Flower arrangement: solitary or small racemes in leaf-axils.
Flowering time: June-August.
Leaf: ovate, opposite, toothed, shortly stalked.
Fruit: 4-8 cm long.

GREAT WILLOWHERB
Epilobium hirsutum

This willowherb is tall, up to 150 cm, erect and densely covered in soft hairs. It often forms stands in marshes and on the banks of streams.
Flower: deep purplish pink, 1·5-2·5 cm across; petals 4, notched; sepals 4, erect, pointed; stamens 8; stigma cream-coloured.
Flower arrangement: raceme, flower stalks arising from leaf-axils.
Flowering time: July-August.
Leaf: rather narrow, opposite, pointed, teeth curved forward; stalkless, leaf base weakly clasping stem.
Fruit: 5-8 cm long.

ROSEBAY WILLOWHERB
Epilobium angustifolium

The tall, almost hairless plants are up to 120 cm high. In Britain a century ago this species was confined to certain localities but has since spread widely. It has a preference for burnt ground, and covered bombed sites in World War II.
Flower: rosy-purple, 2-3 cm across; petals 4, upper 2 slightly larger; sepals 4, narrow, purple; stamens 8.
Flower arrangement: raceme, long, many-flowered.
Flowering time: July-September.
Leaf: narrow, pointed, alternate, 5-20 cm long, entire or with small, widely spaced teeth.
Fruit: 2·5-8 cm long.

BOGBEAN FAMILY Menyanthaceae

BOGBEAN
Menyanthes trifoliata

The leaves and beautiful fringed, pink and white flowers of this hairless water plant are borne above the surface of shallow water, the stems creeping through the mud. It is found in marshes, lake-edges and upland pools.
Flower: pink, paler within, about 1·5 cm across; petals 5, covered in long, white hairs, joined at base; anthers reddish.
Flower arrangement: clustered, on long stalk.
Flowering time: May-July.
Leaf: of 3 rounded leaflets, 3·5-7 cm, alternate, on stalk with sheathing base.
Fruit: capsule, spherical, with persistent style.
Uses: the underground stem contains a bitter substance called menyanthin, which was used as a tonic and to bring down fever.

MISTLETOE FAMILY Loranthaceae

Most of the 1,500 or so species in this family are semi-parasitic shrubs – they contain chlorophyll but rely on various other plants for their water and some of their food, which they obtain by sending suckers into the host root or stem.

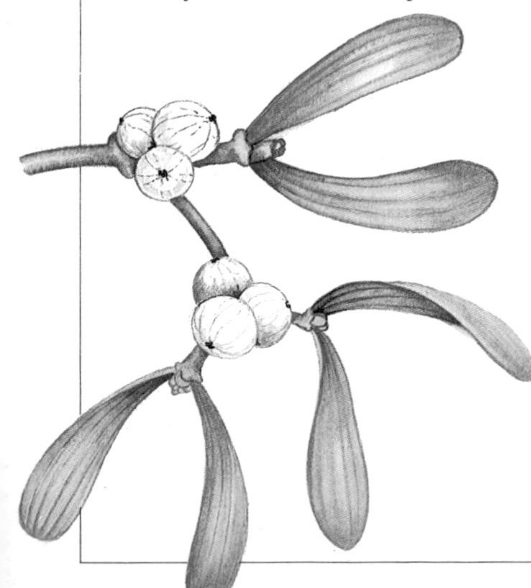

MISTLETOE
Viscum album

Well-known as a Christmas decoration, this much-forked shrubby plant grows on a wide variety of deciduous trees, especially apple and poplar: occasionally on conifers. Its tangle of green branches may form a ball as much as 1 m across.
Flower: small and green with 4 petals: male and female flowers on separate plants. February-April.
Leaf: yellowish-green, thick and leathery: elongated oval to 8 cm long. In pairs.
Fruit: a white berry with sticky flesh, ripening November-February. Birds eat the flesh and then wipe their beaks on the tree branches, leaving the seeds glued to the bark and ready to grow.

STARWORT FAMILY Callitrichaceae

WATER-STARWORT
Callitriche stagnalis

This small, hairless water plant may be found creeping and rooting on wet mud, or floating or submerged in ponds, streams or lakes. The stems grow 1-25 cm long on land and to 100 cm in water. The species is extremely variable and difficult to separate from several close relatives. Leaf shape depends on whether the plants are submerged or not, and on water depth and movement.

Flower: inconspicuous, green, without petals, unisexual.
Flower arrangement: in leaf-axils.
Flowering time: April-October.
Leaf: oval or strap-shaped, opposite, often forming rosettes at tips of emerging shoots.
Fruit: 4-lobed.

MARE'S TAIL FAMILY Hippuridaceae

MARE'S TAIL
Hippuris vulgaris

Growing 25-75 cm high in still or slow-moving fresh water, the unbranched shoots of Mare's Tail bear many whorls of narrow leaves. The leaves are flattened, not tubular as in the superficially similar Horsetails, which are relatives of the ferns. The Mare's Tail is the only member of its family.

Flower: inconspicuous, without petals, greenish, wind-pollinated.
Flower arrangement: in the leaf-axils of aerial shoots.
Flowering time: June-July.
Leaf: very narrow, 1-7·5 cm long, in whorls, stalkless; submerged leaves very thin, pale, limp.
Fruit: nut, green, 2-3 mm.

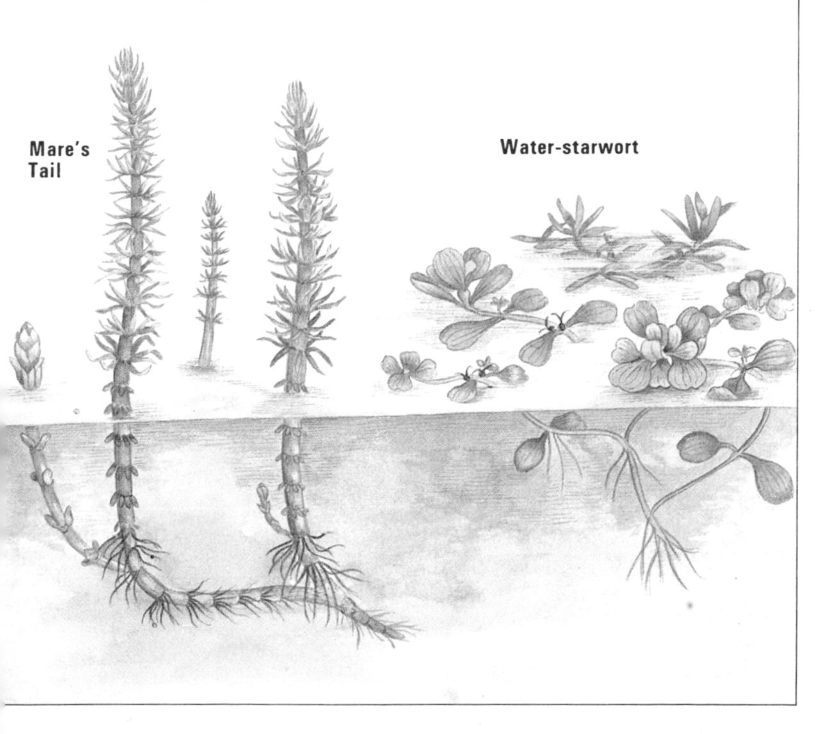

IVY FAMILY Araliaceae

This is a family of mainly tropical herbs, shrubs and trees. The small flowers are green or white, and the fruits are berry-like. Climbing species often have special aerial roots which cling to tree trunks and other surfaces.

IVY
Hedera helix

The familiar, evergreen Ivy climbs to 30 m on trees, cliffs and buildings. Short, thick roots up one side of the stem help it to cling to a support. It may also be found carpeting the ground in woods.
Flower: yellow-green; petals 5, 3-4 mm long, pointed; calyx with 5 tiny teeth; stamens 5; style surrounded by nectar-secreting disc.
Flower arrangement: in rounded, umbel-like clusters on unshaded, high parts of plant.
Flowering time: September-November. The nectar-rich flowers are eagerly sought by insects, especially young wasp queens stocking up their bodies for the winter.
Leaf: glossy, stalked, of 2 kinds: those on non-flowering branches 3-5-lobed, those on the flowering not lobed, ovate or more or less diamond-shaped.
Fruit: black, berry-like, 8-10 mm.

CARROT FAMILY Umbelliferae

Among the most easily recognizable features of this large family is the arrangement of tiny flowers in umbels. These are flat-topped clusters of flowers, with the bases of the stalks (*rays*) all joining at one point. Most species have compound umbels. These are umbels formed, in turn, of smaller umbels. Bracts may be present at the base of a compound (*primary*) umbel and smaller bracts (*bracteoles*) at the bases of the individual (*secondary*) umbels. The fruits and bracts are important for identification.
There are 5 separate petals in each flower, often of markedly different sizes. There may be 5 sepals, but they are often absent. The massed heads attract plenty of insects to the freely exposed nectar.
There are many herbs and spices in the family, including parsley, fennel and dill, as well as many poisonous plants.

WILD CARROT
Daucus carota

The Wild Carrot usually has stiffly hairy stems, ridged, not hollow and 30-100 cm high. It grows in grassy places, and at the coast. It is a biennial, flowering and fruiting in its second year. The widely cultivated carrot with a swollen edible tap root is a form of the Wild Carrot.
Flower: white, middle flower of umbel often red; petals 5, tips incurved; petals of outer flowers enlarged.
Flower arrangement: compound umbel, 3-7 cm across; rays many; bracts conspicuous, each of 3 or more long points; bracteoles similar.
Flowering time: June-August.
Leaf: finely dissected, often fleshy if growing by the sea.
Fruit: ovoid; ridges spiny; 2-4 mm; umbel becoming hollow in fruit.

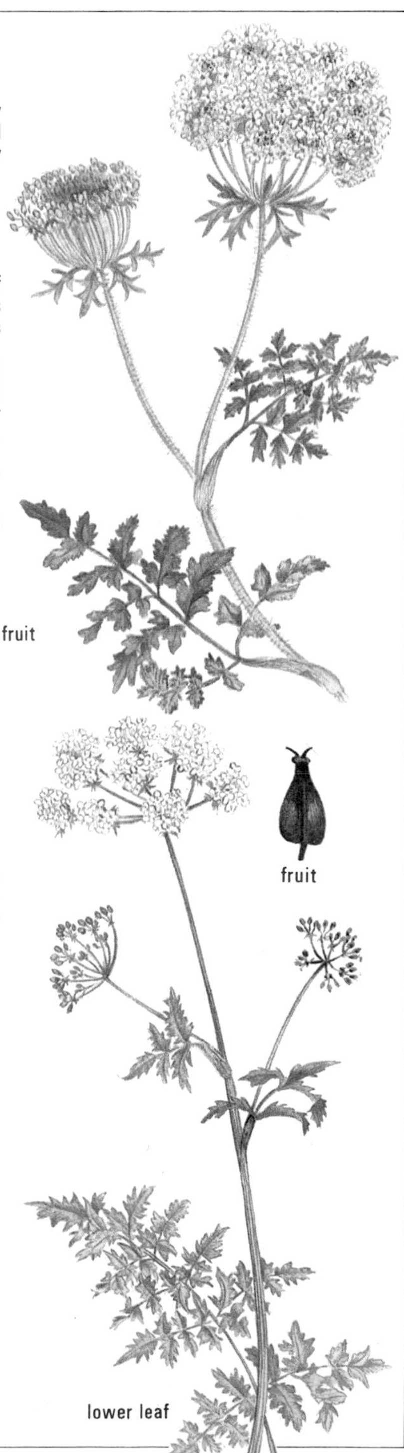

COW PARSLEY
Anthriscus sylvestris

One of the commonest white-flowered umbels, Cow Parsley is also among the earliest to flower. A tall, slightly downy plant, it grows up to 150 cm high in hedges and wood borders. The stems are hollow, grooved and may be tinged red.
Flower: white, 3-4 mm across; petals 5, tip often slightly notched and curved inward.
Flower arrangement: in compound umbels, 2-6 cm across; no bracts; bracteoles 5 or 6.
Flowering time: April-June.
Leaf: 2- or 3-pinnate, up to 30 cm long, shortly hairy beneath, leaf-stalk sheathing stem at base.
Fruit: of 2 oblong, joined carpels, black, smooth, 5-10 mm, each bearing short, persistent style.

FOOL'S PARSLEY
Aethusa cynapium

This common weed of gardens, field margins, and waste places is a rather shiny, bright green annual. It is quite hairless and the stems are very smooth – pale green with prominent darker striations. Reaching 120 cm, it is a poisonous plant and its vague resemblance to cultivated parsley has given it its name.

Flower: white, 2 mm across with very unequal petals. Borne in compound umbels 2-6 cm across. Bracts are usually absent, but slender, strap-shaped bracteoles about 1 cm long hang from the secondary umbels like beards and make this plant easily identified when in flower. June-October.

Leaf: triangular in outline and 2-3 pinnate; with membranous flaps where leaf bases clasp the stems.

Fruit: egg-shaped and clearly ribbed: not spiny.

WILD ANGELICA
Angelica sylvestris

Often stout and up to 200 cm or more high, this Umbellifer is almost hairless. It favours damp, shady places. The hollow, grooved stem is often purple with a whitish cast or bloom.

Flower: white or pink, 2 mm across; petals 5, incurved.

Flower arrangement: compound umbel, rounded, 3-15 cm across; rays many, slightly hairy; few or no bracts; bracteoles few, bristle-like.

Flowering time: July-September.

Leaf: doubly or trebly pinnate; leaflets 2-8 cm, ovate, toothed, leaf-stalks deeply grooved on upper side, widely sheathing stem at base. Upper leaves are little more than sheaths with a tuft of leaflets at the tip, often with small umbels in axils.

Fruit: ovoid, 4-5 mm, flattened, winged; persistent styles curved.

HEMLOCK
Conium maculatum

Hemlock is extremely poisonous in all parts. It can be recognized by the smooth, purple-spotted stem and a 'mousy' smell if the plant is crushed. The hairless plant grows up to 2·5 m in damp habitats, near water and on waste land. It is a biennial.
Flower: white, 2 mm across; petals 5, tips shortly incurved.
Flower arrangement: compound umbel, 2-5 cm across; rays 10-20; bracts 5-6; bracteoles on outer sides of secondary umbels.
Flowering time: June-July.
Leaf: finely dissected, up to 30 cm.
Fruit: spherical, 2·5-3·5 mm, ridges bumpy.
Uses: traditionally this was the drug used to poison Socrates. Medicinal use of Hemlock was revived in the late 18th century but discontinued due to the uncertain effects.

GROUND ELDER
Aegopodium podagraria

The leaves of Ground Elder have broad, undissected leaflets, unlike the finely dissected leaves of many other Umbellifers. The plant is hairless, with hollow, grooved stems, and grows in shady places and gardens, where it can be a stubborn weed.
Flower: white, sometimes pink, 1 mm across; petals 5, tips incurved.
Flower arrangement: in compound umbels, 2-6 cm across; rays 15-20; few or no bracts or bracteoles.
Flowering time: May-July.
Leaf: of 3 leaflets, each 4-8 cm, ovate, toothed, on long, 3-angled leaf-stalk sheathing stem at base.
Fruit: ovoid, 4 mm, ridged; persistent styles curved down.

LESSER WATER-PARSNIP
Berula erecta

These hairless plants, 30-100 cm high, grow in shallow fresh water, often forming sprawling masses. The stems are hollow and ridged. Fool's Water-cress (*Apium nodiflorum*) is similar to this species. They cannot easily be told apart unless flowers or fruit are present.
Flower: white, 2 mm across; petals 5, tips incurved.
Flower arrangement: compound umbel, 3-6 cm across, on long stalk; rays 10-20; bracts and bracteoles many, often with a few teeth.
Flowering time: July-September.
Leaf: pinnate, up to 30 cm long; 7-14 pairs ovate, toothed leaflets; dull, blue-green.
Fruit: 2-lobed, 1·5-2 mm, a little wider than high; persistent styles curved. A poisonous species.

HOGWEED
Heracleum sphondylium

The stout stem of Hogweed is grooved, hollow and stiffly hairy, the hairs bent sharply down. The plant is found in grassland, hedges and woods, growing 50-200 cm high. The Giant Hogweed (*H. mantegazzianum*) is up to 5 m and has a red-blotched stem up to 10 cm thick.
Flower: white or pink, 5-10 mm across; petals 5, incurved, enlarged in outer flowers.
Flower arrangement: compound umbel, 5-15 cm across; rays 7-20, few or no bracts; bracteoles bristle-like.
Flowering time: June-September.
Leaf: pinnate, 15-60 cm; leaflets broad, toothed and lobed, stiffly hairy; bases of stalks widely sheathing stem.
Fruit: ovoid, flattened, whitish, 7-8 mm; persistent styles short, curved.
Uses: the leaves are fed to pigs.

BURNET SAXIFRAGE
Pimpinella saxifraga

This species has a downy and tough but slender stem, weakly ridged. It grows 30-100 cm high in dry grassland, very often on chalk or limestone. The seed of a Mediterranean species of *Pimpinella* is the aniseed of seed cake and sweets.
Flower: white, 2 mm across; petals 5, tips incurved.
Flower arrangement: compound umbel, 2-5 cm across, flat-topped; rays 10-20; no bracts or bracteoles.
Flowering time: July-August.
Leaf: pinnate; leaflets ovate, coarsely toothed, stalkless; upper leaves doubly pinnate, leaflets narrow.
Fruit: ovoid, 2-3 mm.
The Greater Burnet Saxifrage (*P. major*) is a taller, hairless relative, often with pink flowers (especially in the north). All leaves are once pinnate. It grows on heavier and damper soils.

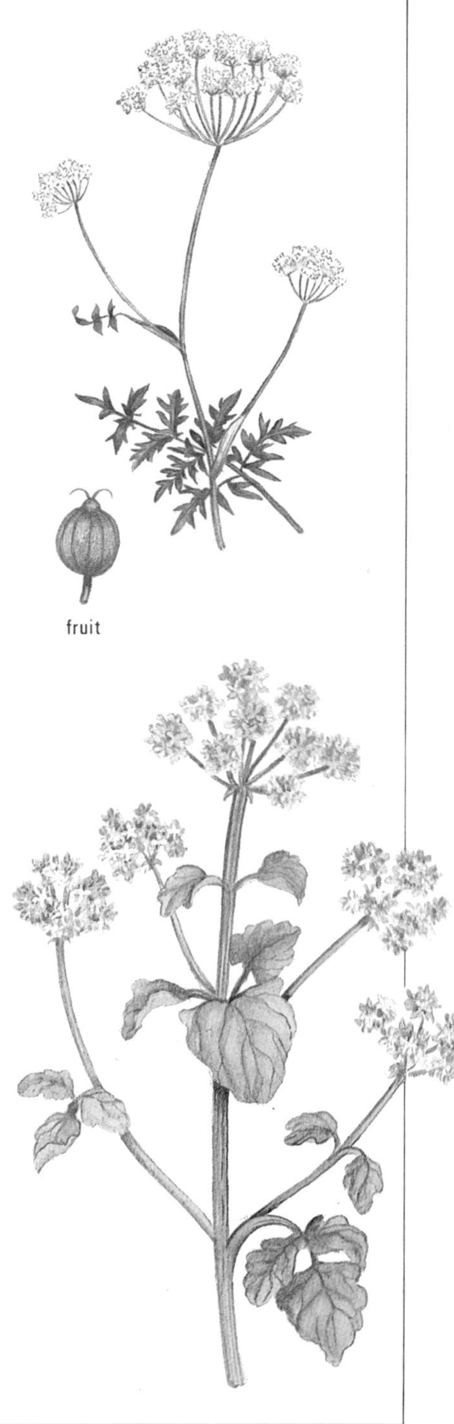

ALEXANDERS
Smyrnium olusatrum

This tall and rather imposing biennial grows mainly in coastal regions. Its stout branched stems and shiny foliage form dense banks on roadsides, cliffs, and waste ground. It reaches 150 cm and has a distinct smell of celery. A native of the southern half of Europe, it is now well established in Britain. It was once cultivated as a pot herb.
Flower: yellow, 1-5 mm across in domed or globular compound umbels with few bracts or bracteoles. April-June.
Leaf: bright green, often yellowish when young. Basal leaves 30 cm long or more and much divided (thrice trifoliate); leaflets – diamond-shaped. Upper leaves twice trifoliate.
Fruit: black and globular, about 8 mm long.

SANICLE
Sanicula europaea

This hairless plant grows 20-60 cm high and forms dense carpets in many woodlands, especially beechwoods.
Flower: greenish white, often with a pinkish tinge; very small and packed into small globular umbels, 3 or 4 of which are in turn grouped into loose compound umbels. Bracts 3-5 mm long and pinnately lobed: bracteoles slender and undivided. May-August.
Leaf: deep green and shiny with reddish stalks; palmately lobed.
Fruit: oval, 3 mm long and clothed with hooked bristles.

SEA HOLLY
Eryngium maritimum

An inhabitant of sand dunes and shingle banks by the sea, this beautiful plant forms a blue haze over the ground when in flower. It reaches 60 cm, but is often shorter.
Flower: bright blue and packed into a simple spiny umbel which is oval and up to 4 cm long. The umbel is surrounded by broad and spiny leaf-like bracts. July-September.
Leaf: grey-green with a waxy coating and sharply spined edges reminiscent of true holly. The basal leaves are often markedly fan-shaped.
Fruit: oblong, about 5 mm long and covered with hooked bristles.

SPURGE FAMILY Euphorbiaceae

This large, mainly tropical family includes plants which produce rubber and castor oil. Many are trees and shrubs, but all the European species are herbaceous. Spurges have very distinctive flowers without petals. A typical inflorescence consists of a little cup in which there are several minute, one-stamened male flowers and a single female flower with three stigmas. Nectar-secreting glands around the edge attract insects. All members of the family contain a poisonous and often milky sap.

SUN SPURGE
Euphorbia helioscopia

The whole plant is hairless, 10-50 cm high and erect. The characteristic yellowish bracts of the flowers lend a yellow-green look to the plant. The stem is often red. An annual living on disturbed ground.
Flower: minute; glands on rim green; kidney-shaped, surrounded by yellowish, leaf-like bracts.
Flower arrangement: in umbel-like, 5-stalked clusters.
Flowering time: May-October.
Leaf: rounded, lower leaf often tapering at base; minutely toothed.
Fruit: capsule, smooth, 3-5 mm, stalked, nodding over rim of flower.

PETTY SPURGE
Euphorbia peplus

Growing 10-30 cm high in cultivated ground, this annual is hairless and light green. The minute flowers are surrounded by leaf-like bracts.
Flower: minute; glands on rim crescent-shaped with long, pointed tips, surrounded by opposite, leaf-like bracts.
Flower arrangement: flower stalks repeatedly dividing in two.
Flowering time: April-November.
Leaf: oval, not toothed, alternate.
Fruit: capsule, ridged, 2 mm, stalked, nodding over rim of flower.

DOG'S MERCURY
Mercurialis perennis

Poisonous to humans and livestock, Dog's Mercury is erect, up to 40 cm, unbranched and shortly hairy. The plant smells unpleasant if crushed, and grows in woods and shady places.
Flower: inconspicuous, green; male and female on separate plants.
Flower arrangement: males in slender spikes, females in small, stalked clusters.
Flowering time: February-April.
Leaf: elliptic, blunt-toothed, opposite, 2-8 cm long, shortly stalked.
Fruit: spherical, hairy, 6-8 mm.

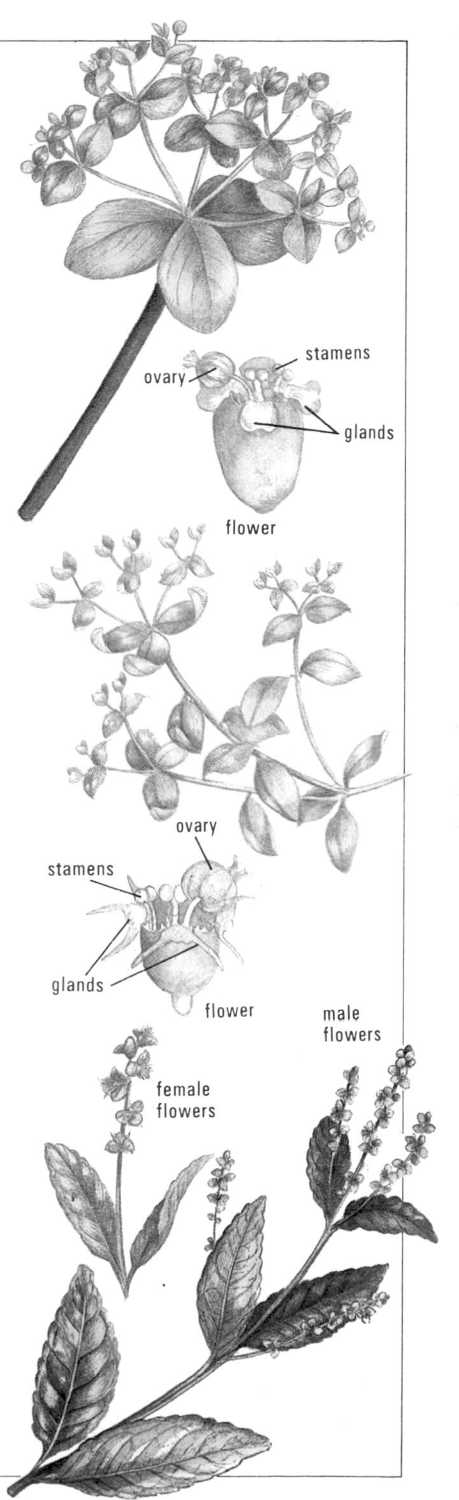

GOURD FAMILY Cucurbitaceae

Most members of this family are tropical climbing herbs. They grow rapidly and climb by means of spring-like tendrils. There are usually 5 sepals and 5 petals. Well-known members of the family include melons, marrows, and cucumbers – all with thick-skinned berries as fruit.

WHITE BRYONY
Bryonia dioica

This rather dull green climber is abundant in hedgerows and woodland margins, where its bristly angular stems readily reach heights of 4 m. Its tendrils coil in two different directions and their great elasticity allows the plant to sway in the wind without risk of breaking away from its support.
Flower: in small clusters, male and female on separate plants: pale green with hairy petals. Female flowers 10-12 mm across, male flowers 12-18 mm. May-September.
Leaf: hairy and palmately lobed with 3-5 toothed lobes. The unbranched tendrils arise close to the leaf stalks and may be modified stipules, although their exact nature is not clear.
Fruit: a red berry. Poisonous. The plant is in no way related to the Black Bryony (page 129) which often grows with it.

THRIFT FAMILY Plumbaginaceae

Many members of this family grow on the coast. The remainder are alpine, Arctic or desert species. The flowers have 5 petals and papery sepals.

THRIFT
Armeria maritima

Thrift forms cushions of narrow leaves and pink flowers mainly in rocky places at the seaside, in salt marshes, or inland on mountains and sandy places.
Flower: pale to deep pink; petals 5.
Flower arrangement: in dense, rounded head, 1·5-2·5 cm across; bracts below, greenish; flower stalk 5-30 cm high, surrounded by brown, papery sheath above.
Flowering time: April-October.
Leaf: grass-like, slightly fleshy, hairy or not, 2-15 cm long, up to about 4 mm wide.
Fruit: surrounded by papery calyx.

COMMON SEA LAVENDER
Limonium vulgare

This is a saltmarsh plant growing to about 40 cm, often turning the marshes blue with flowers in summer.
Flower: up to 8 mm across, with a papery pale lilac calyx and 5 purplish-blue petals: densely clustered on arching spikes to form flat-topped heads. July-August. Calyx persists after flowering.
Leaf: all from the base; elliptical to 12 cm long; hairless.

HEMP FAMILY Cannabiaceae

There are two main divisions in this family. One contains the Hops. The other contains species used for their plant fibre (hemp) and the drug cannabis. The flowers are either male or female, borne on separate plants.

HOP
Humulus lupulus

The Hop is found climbing with rough stems in hedges and, although native in N. Europe, is often an escape from cultivation. The 'cones' of pale bracts on the female plants are easy to recognize.
Flower: males and females on separate plants; females enclosed by yellow-green bracts; males tiny, green; petal-like lobes 5.
Flower arrangement: female bracts in hanging 'cone'; male flowers in branched clusters.
Flowering time: July-August.
Leaf: 3-5 lobed, coarsely toothed, roughly hairy, opposite, 10-15 cm, yellow glands dotting underside.
Fruit: enclosed by enlarged bracts dotted with yellow resin glands.
Uses: the female 'cones' are used to flavour beers. Cultivation of Hops began in Britain in about 1520, earlier in the rest of Europe.

DOCK FAMILY Polygonaceae

Most members of this large family are from the northern temperate zone. Useful species include buckwheat and rhubarb. The leaves have a characteristic papery sheath called an ochrea encircling the stem at the leaf-base. The small flowers have no true petals, but the sepals are white, green, or pink and petal-like. The fruit, a small nut enclosed by the persistent flower, is important in identification.

KNOTGRASS
Polygonum aviculare

Knotgrass is a ragged-looking annual, erect (up to 2 m high) or straggling in mats on trampled ground and at the coast. The plant is hairless and branched, and the ochreae are ragged and silvery.
Flower: pink or white; petal-like lobes 5.
Flower arrangement: 1-6 flowers in the leaf-axils.
Flowering time: July-October.
Leaf: elliptic to very narrow, tips more or less pointed, almost stalkless; leaves on main stem larger than branch leaves; ochreae silvery, torn.
Fruit: nut, 3-angled, hidden by dead flower.

REDSHANK
Polygonum persicaria

The stem, often reddish, is more or less erect, branched and 25-80 cm high. There are dense spikes of tiny pink flowers, and the leaves often bear a dark blotch. The whole plant is almost hairless and grows on waste and cultivated ground. It is an annual.
Flower: pink; petal-like lobes 5.
Flower arrangement: compact spike, up to 3·5 cm long.
Flowering time: June-October.
Leaf: narrow, tapering to point, often dark-blotched; ochreae brownish, fringed with hairs.
Fruit: nut, black, shiny.

BLACK BINDWEED
Polygonum convolvulus

The grooved stem of the climbing Black Bindweed twines clockwise around its support to a height of 30-120 cm. It is an annual growing on waste and cultivated land.
Flower: greenish white; petal-like lobes 5, 3 outer with ridge; each flower stalk 1-2 mm.
Flower arrangement: loose spike or raceme in the leaf-axils.
Flowering time: July-October.
Leaf: spear-shaped, stalked, powdery white beneath, usually alternate; ochrea rim at an angle.
Fruit: nut, dull, black.

AMPHIBIOUS BISTORT
Polygonum amphibium

The stems and leaves of this water plant float on the surface of still or slow-moving water. The stems grow 30-75 cm long and root along their length. A slightly differing form, with short, stiff hairs on the leaves and ochreae, grows at the water's-edge and in damp grassland.
Flower: pink; petal-like lobes 5; stamens 5, red.
Flower arrangement: dense, short spike, 2-4 cm long, stalked.
Flowering time: July-September.
Leaf: long—ovate, hairless, shiny stalked, alternate.
Fruit: nut, shiny, brown.

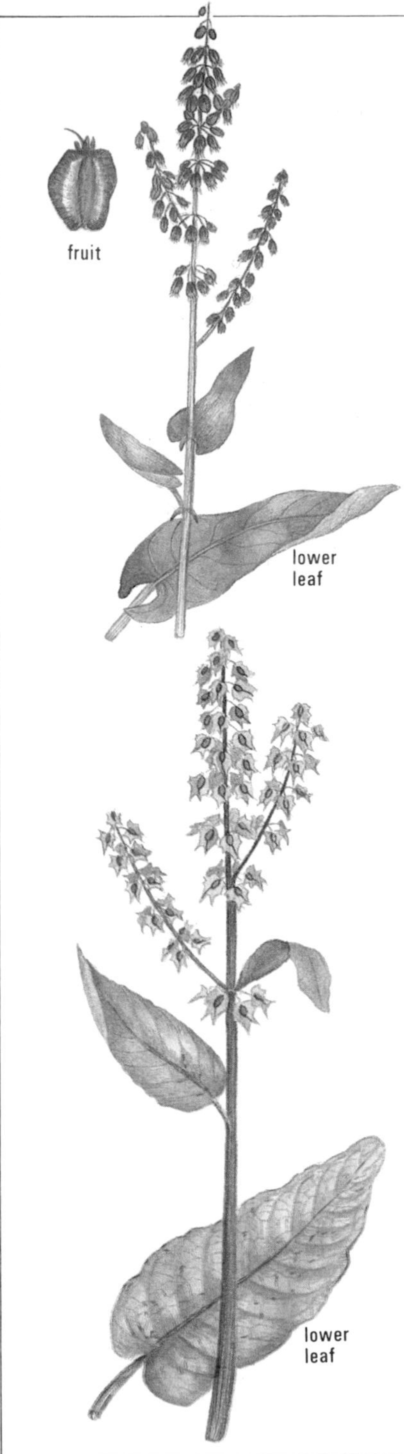

COMMON SORREL
Rumex acetosa

Growing up to 100 cm but usually much smaller, this almost hairless plant grows in grassland, to which the flowers and fruits can give a reddish tinge.
Flower: green and red, tiny; male and female flowers on separate plants; wind-pollinated.
Flower arrangement: in whorls on branched raceme.
Flowering time: May-July.
Leaf: spear-shaped, basal lobes pointing down, acid-tasting, up to 10 cm long; upper leaves almost stalkless; ochreae fringed.
Fruit: 3-4 mm, with 3 papery wings, tinged red, enclosing 3-angled nut.
Uses: a sharp-tasting sauce, to be eaten with fish or pork, can be made from the leaves simmered in butter with salt and black pepper.

BROAD-LEAVED DOCK
Rumex obtusifolia

This is a very common dock, 50-120 cm high, of waste places and cultivated land. There are several similar species of dock which can be distinguished by their fruits.
Flower: green, tiny, wind-pollinated.
Flower arrangement: in whorls, on branched raceme.
Flowering time: June-October.
Leaf: oblong, broad, tip pointed or rounded, lobed at base, alternate, up to 25 cm long.
Fruit: 5-6 mm, with 3 green, deeply toothed wings alternating with 3 unequal red swellings enclosing 3-angled nut.
Uses: traditionally dock leaves are rubbed on nettle stings to reduce irritation.

NETTLE FAMILY Urticaceae

Stinging hairs crop up in many members of this family. The flowers are inconspicuous and without petals. The stamens are flicked out as the flowers open, expelling pollen in puffs.

COMMON NETTLE
Urtica dioica

Notoriously covered in stiff, stinging hairs, the Common Nettle grows 30-150 cm high in all kinds of disturbed places. When touched, the hair-tip breaks off. The remaining hollow hair injects the skin with fluid containing an irritant poison.
Flower: green, tiny; male and female flowers on separate plants in loose-hanging spikes; wind-pollinated.
Flowering time: June-August.
Leaf: ovate, toothed, opposite, 4-8 cm long.
Uses: In World War II, the tough plant fibres were made into textiles.

HEATHER FAMILY Ericaceae

As well as heathers and heaths, this family also contains the rhododendrons. The members are small or large shrubs, mainly from temperate areas, with evergreen, often needle-like leaves. The flowers are normally bell-shaped. Most species prefer acid soils, and many have a fungus growing inside the roots.

HEATHER
Calluna vulgaris

This low-growing, wiry shrub grows 15-80 cm high, preferring acid soil and carpeting large tracts of heath, moor and open woods. Heather differs from the similar species of Heath in that the flower petals are not joined into a bell and the leaves are opposite, not whorled.
Flower: mauve, scented; petals 4, hidden by 4 petal-like sepals, about 4 mm long.
Flower arrangement: spikes, 3-15 cm.
Flowering time: July-September.
Leaf: needle-like, opposite, pressed to stem, 1-3·5 mm long, overlapping on short, side shoots.
Fruit: capsule.
Uses: hives of bees are kept on Heather moors in late summer, the resulting honey being particularly good. Heather also feeds sheep and shelters red grouse.

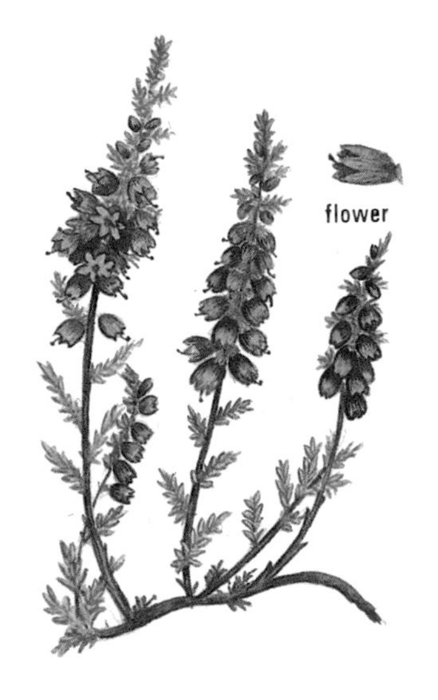

BELL HEATHER
Erica cinerea

Often confused with the previous species, this sprawling under-shrub is abundant on dry heathland in western Europe. The upright stems reach 60 cm and are clothed with short leafy shoots. Less branched and contorted than Heather.

Flower: bell or urn-shaped, 5-6 mm long; deep pink or purple with dark green sepals. In short racemes. July-September.

Leaf: bristle-like, dark green and hairless; 3-6 mm long. In whorls of 3 on the shoots.

Fruit: a many-seeded capsule.

ALPENROSE
Rhododendron ferrugineum

This is a low-growing evergreen shrub of mountain slopes, to an altitude of 3,200 m in the Alps and Pyrenees, mainly on acidic rocks. It grows to a height of 1 m and forms neat, rounded bushes on windswept scree slopes. The bushes may join up to form extensive thickets. The plant also carpets open woodland on the mountains.

Flower: pale pink to deep red, bell-shaped or funnel-shaped about 15 mm long; sepals very short. Borne in small clusters at tips of the branches.

Leaf: elliptic or oblong, 3-5 cm long: dark green above and clothed with reddish-brown scales beneath: margins rolled under.

Fruit: a dry capsule.

The Hairy Alpenrose (*R. hirsutum*), confined to the Alps, is similar but lacks the brown scales under the leaves.

CROSS-LEAVED HEATH
Erica tetralix

This small undershrub is closely related to Bell Heather, but is rarely more than 30 cm high and is less branched. It is a downy plant and favours wet heaths and bogs. It is absent from the drier heathlands.
Flower: pale pink, 6-7 mm long and flask-shaped with a narrow opening; sepals short, green and downy. Borne in small drooping clusters at tips of shoots.
Leaf: greyish and downy, tightly rolled; 2-5 mm long. In whorls of 4 in the shape of a cross.

BILBERRY
Vaccinium myrtillus

A low, hairless shrub 15-35 cm high, the Bilberry is known for its sweet, black berries. It grows in woods, and on moors and mountains to an altitude of over 1,200 m. The twigs are 3-angled and green.
Flower: pink, tinged green; petals joined into a globe shape, 4-6 mm.
Flower arrangement: solitary or paired.
Flowering time: April-June.
Leaf: ovate, tip pointed, minutely toothed, bright green, short-stalked, alternate, 1-3 cm long.
Fruit: berry, blue-black, edible.
Uses: the berries are worth the labour of gathering. If too bitter for eating raw then they make good tarts and jam.

CRANBERRY
Vaccinium oxycoccus

This prostrate, creeping evergreen undershrub, a relative of the previous species, grows in acidic bogs, where its thread-like stems creep over the peat and root at intervals.
Flower: drooping, with 4 deep pink petals about 5 mm long and strongly reflexed, leaving the cluster of dark stamens projecting downwards like a beak. Borne singly or in pairs on slender reddish stalks up to 3 cm. June-August.
Leaf: oval, 4-10 mm long, carried alternately on the creeping stems. Dark green above, waxy grey beneath, with margins rolled under.
Fruit: a spherical or pear-shaped berry, 6-8 mm across, red or brown and usually heavily speckled. It is unpalatable when raw but widely used in preserves.

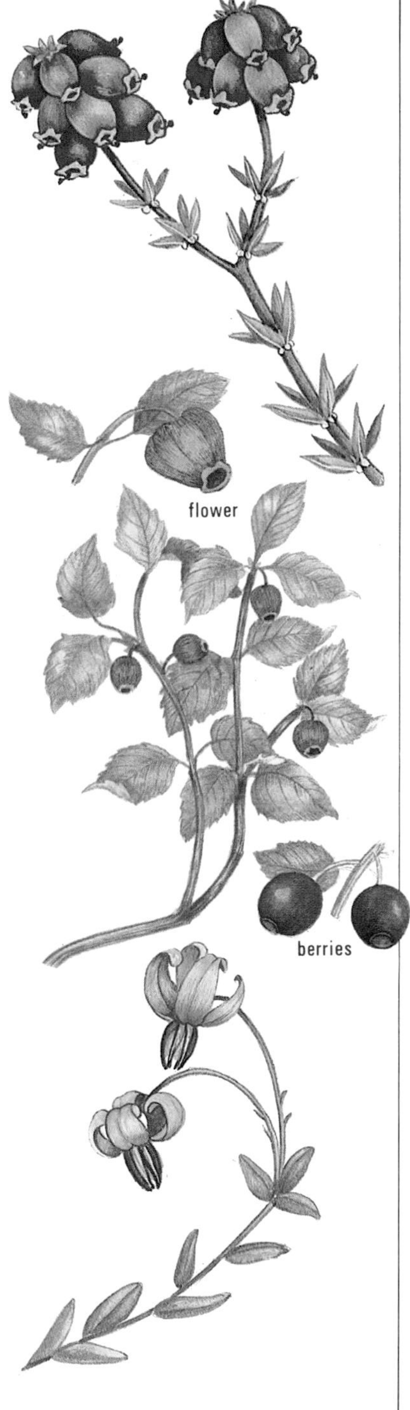

PRIMROSE FAMILY Primulaceae

This is a large family in which the petals are joined together and often form a tube, as in the primroses themselves. Many species have two kinds of flowers – *pin-eyed*, in which the stigma shows in the throat of the tube with the stamens far below it, and *thrum-eyed* with the stamens in the throat and the stigma below them. Such an arrangement encourages cross-pollination.

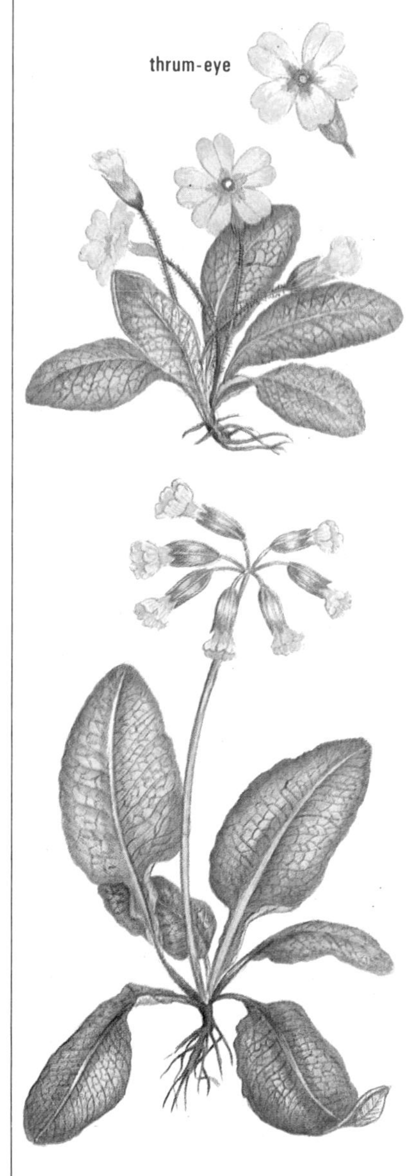

PRIMROSE
Primula vulgaris

This much-loved plant grows in grassy places, woods and hedges, flowering before the trees come into leaf and shade the ground.
Flower: pale yellow, deep yellow in throat, 2-3 cm across; petals joined at bases into tube with 5 spreading, notched lobes; calyx tubular, 5-toothed.
Flower arrangement: solitary on long, softly hairy stalks.
Flowering time: December-May.
Leaf: oblong, tip rounded, tapering gradually at base into stalk, wrinkled, shortly hairy beneath.
Fruit: capsule, ovoid, enclosed by calyx.

COWSLIP
Primula veris

Cowslip is only locally common and is decreasing owing to the ploughing of old pasture where it grows. The plants are 5-30 cm high.
Flower: pale or deep yellow, orange in throat, nodding; petals joined at bases into tube with 5 notched lobes; calyx tubular, pale green, 5-toothed.
Flower arrangement: 1-30 drooping flowers in umbel-like cluster on shortly hairy stalk.
Flowering time: April-May.
Leaf: oblong, rounded at tip, abruptly narrowed into stalk, wrinkled, shortly hairy all over.
Fruit: capsule, ovoid, hidden by calyx.

BIRD'S-EYE PRIMROSE
Primula farinosa

This beautiful little plant grows in moist meadows in upland regions of northern England and Scandinavia and on damp mountain slopes elsewhere in Europe. It grows to 20 cm.
Flower: pink or purple, occasionally white, with a yellow eye; about 1 cm across. Grouped into a loose umbel at the top of a mealy white stem (mealiness may disappear from older stems). May-July.
Leaf: spatula-shaped, 1-5 cm long, lightly toothed; light green above, densely mealy white below.
Fruit: a small capsule.

SCARLET PIMPERNEL
Anagallis arvensis

One of the very few scarlet-flowered plants native to N. Europe, the Scarlet Pimpernel reaches 30 cm on dry grassland and as an annual weed on cultivated ground.
Flower: scarlet or salmon pink, rarely blue, up to 14 mm across; petals 5, more or less rounded; sepals 5, sharply pointed.
Flower arrangement: on thin stalks, longer than leaves, arising from each leaf-axil.
Flowering time: June-August.
Leaf: ovate, pointed, opposite, stalkless, black glands beneath.
Fruit: capsule, spherical, with persistent style, splitting around middle.
Uses: the flowers can be used to tell the time and approaching weather. They open from about 8 am to 3 pm and close in the damp, cool air of impending rain.

seed capsule

YELLOW LOOSESTRIFE
Lysimachia vulgaris

The tall, erect, shortly hairy plants of Yellow Loosestrife grow 60-150 cm high. They are found along riversides and other damp places. This species is absent from N. Scotland, Sweden and much of Norway.
Flower: yellow, about 1·5 cm across; petals 5; sepals 5, orange-margined.
Flower arrangement: in loose, branched heads.
Flowering time: July-August.
Leaf: long, narrow, pointed, in pairs or whorls of 3 or 4; stalkless, dotted with black glands.
Fruit: capsule, spherical, opening by 5 valves.

PERIWINKLE FAMILY Apocynaceae

Members of this large family are mainly tall rain-forest trees of the tropics. The rest are shrubs and woody climbers, or herbs in northern Europe, with a milky, often poisonous latex, used in the tropics to tip arrows for hunting.

Lesser Periwinkle

LESSER PERIWINKLE
Vinca minor

This species has long been grown in gardens. It has escaped and become naturalized in shady places but is truly native in many parts of Europe, where it is common on the floor of many woodlands. Probably native in Britain only in the south.

It is an evergreen plant with long, rooting stems of 30-60 cm covering the ground.

Flower: blue-purple or white, 2·5-3 cm across; petals 5, wide and blunt at tip, joined at base; calyx tube 5-toothed, hairless.

Flower arrangement: solitary, stalked, in leaf-axil.

Flowering time: March-May.

Leaf: elliptic, opposite, glossy, hairless, short-stalked.

Fruit: pair of follicles, rarely ripening in Britain, forked at tip.

The Greater Periwinkle (*V. major*) is very similar but its trailing stems root only at their tips. The flowers are 4-5 cm across with fringed sepals: up to 4 flowers at each node of the stem.

GENTIAN FAMILY Gentianaceae

Many of the species in this family grow in arctic regions and on mountains. They are nearly all herbaceous plants and their leaves are quite hairless. The flowers, of 5 joined petals, are often blue, bell-shaped or trumpet-shaped, and erect. The Gentian family is one in which a fungus commonly lives within the roots, benefitting both plant and fungus. The bitter roots have been much used medicinally.

TRUMPET GENTIAN
Gentiana acaulis

This is one of the most striking gentians and a favourite species for the rock garden. It grows in the Alps, Pyrenees, and Apennines, reaching altitudes of 3,000 m. Preferring acidic rocks, it grows in turf and stony places, especially where there is plenty of moisture.

Flower: a deep, erect trumpet up to 7 cm tall, with 5 spreading, pointed lobes. Deep, bright blue with patches of green spots around the throat. Carried singly on leafy stalks up to 5 cm long. May-August.

Leaf: dark and slightly greyish green; elliptical to 15 cm long: in a loose rosette.

SPRING GENTIAN
Gentiana verna

This intensely blue gentian, one of several similar species, is found mainly on the mountains, although it occurs (rarely) at lower altitudes in Britain and Ireland. It prefers calcareous soils, where it displays sheets of blue flowers on the screes and grasslands as soon as the snows melt. Absent from northern Europe.
Flower: deep blue, with a tube up to 2·5 cm long and 5 oval lobes spreading star-like to 2 cm across. Borne singly on stems to 6 cm high. March-August.
Leaf: bright green and leathery: oval to 15 mm long. In a dense rosette, with a few smaller leaves on the flowering stems.

COMMON CENTAURY
Centaurium erythraea

Common Centaury varies between 2 and 50 cm high, with a branched stem bearing flat clusters of pink flowers. It is a hairless plant growing in dry grassland and dunes. It is an annual.
Flower: rose-pink, 1-1·5 cm across; petals joined into tube below; lobes 5, spreading flat; calyx with 5 long, narrow teeth.
Flower arrangement: in more or less flat-topped clusters.
Flowering time: June-October.
Leaf: opposite, basal leaves in rosette; elliptic, tip rounded or pointed, stalkless.
Fruit: capsule.
Uses: Common Centaury has been called Gall of the Earth due to its bitterness. A tonic can be made from an infusion of dried plants.

FORGET-ME-NOT FAMILY Boraginaceae

Members of this family are almost always hairy, often stiffly so. The blue or mauve, sometimes white, flowers are very often pink in bud and arranged in one-sided cymes which uncoil as the flowers open.

WATER FORGET-ME-NOT
Myosotis scorpioides

Growing in damp or wet places, this forget-me-not has a slightly ridged, hairy stem, 15-45 cm high, the base of which often creeps and roots. The important characteristics of this species are the flower size and the notched petal-lobes. Several similar species, annual or perennial, grow on the drier soils of woods and fields.

Flower: sky-blue, yellow in throat, 4-10 mm across; petals joined into tube at base; lobes 5, shallowly notched; calyx 5-toothed with straight hairs; buds pink.
Flower arrangement: cyme, uncoiling as flowers open, without bracts.
Flowering time: May-September.
Leaf: oblong, usually rounded at tip, stalkless, shortly hairy or almost hairless.
Fruit: 4 nutlets, black, shiny, enclosed by calyx.

VIPER'S BUGLOSS
Echium vulgare

An erect, stout biennial, 20-90 cm tall and densely covered in stiff, whitish hairs with swollen bases, Viper's Bugloss has bright blue flowers, pink in bud, forming an exciting colour combination. It grows in dry places and is common on roadsides in many sandy areas.

Flower: bright blue, 1-2 cm; petals joined into tube at base; 5-lobed; calyx 5-toothed; stamens 5, 4 protruding.
Flower arrangement: a panicle of short, curved or coiled cymes.
Flowering time: June-September.
Leaf: lanceolate, tapering gradually to base, up to 15 cm long, stalkless, lowest shortly stalked.
Fruit: 4 nutlets, enclosed by calyx.
Uses: this plant was linked from ancient Greek times with snakes. The nutlet, supposed to resemble a snake's head, was consequently used to treat snake bite.

COMMON COMFREY
Symphytum officinale

Common Comfrey is a branched, stout, stiffly hairy plant of 30-120 cm. The stem has thin prominent ridges, and the plant grows in damp places.
Flower: cream, white, pink or purple, 15-17 mm long; petals joined into tube with 5 short lobes at rim; calyx 5-toothed; style protruding from petal-tube.
Flower arrangement: in nodding cymes, uncoiling as flowers open.
Flowering time: May-June.
Leaf: ovate to lanceolate, tip pointed, bases running down stem in thin, often wavy ridges.
Fruit: 4 nutlets, black, shiny, enclosed by calyx, style persistent.
Uses: in Bavaria the young leaves are fried in batter. Manure water to feed tomato or marrow crops can also be made by soaking Comfrey plants in water for a week. The resulting water is rich in potassium.

BINDWEED FAMILY Convolvulaceae

The flowers in this family are bell- or trumpet-shaped on stems which often climb by twining. The sap is quite often milky.

FIELD BINDWEED
Convolvulus arvensis

The invasive rhizomes make this little creeping or climbing plant a stubborn weed of farm and garden. The stem coils anti-clockwise.
Flower: pink or white, striped darker outside, trumpet-shaped, 1·5-3 cm across, scented; sepals 5.
Flower arrangement: 1-3 flowers on long stalk with 2 small bracts half-way up.
Flowering time: June-September.
Leaf: arrow-shaped, 2-5 cm long, stalked, shortly hairy or not.
Fruit: capsule, spherical.

HEDGE BINDWEED
Calystegia sepium

This climbing bindweed covers such supports as railings and hedges with a blanket of arrow-shaped leaves and large, white trumpet-shaped flowers. These are visited at dusk by hawk moths. The stems grow 1-3 m long, twisting anti-clockwise.
Flower: white, trumpet-shaped, 3-4 cm across, scentless; sepals 5, enclosed by 2 slightly longer bracts.
Flower arrangement: solitary, on long stalks.
Flowering time: July-September.
Leaf: arrow-shaped, up to 15 cm long, stalked, shortly hairy or not.
Fruit: capsule, enclosed by bracts.
Great Bindweed (*C. sylvatica*) is similar but flowers to 7·5 cm across and swollen bracts hide sepals.

COMMON DODDER
Cuscuta epithymum

This is a parasitic annual quite devoid of chlorophyll. A hair-like red stem 0·1 mm thick emerges from the minute seed and creeps over the ground. If it meets a suitable host plant – a wide variety of herbs and small shrubs will do – it begins to climb and soon sends tiny suckers into the host to absorb food. The dodder stem can then grow rapidly and branch repeatedly until it may completely smother the host, although the stems do not get much thicker. The plant is most common on heathland.
Flower: minute and bell-shaped; pale pink and scented. In dense clusters up to 1 cm across. July-September.
Leaf: reduced to minute red scales on the stem.
Greater Dodder (*C. europaea*) is a little larger (stems 1 mm thick) and usually found on stinging nettles.

POTATO FAMILY Solanaceae

Most of the 2,000 or so species in this family come from the tropics. They include many important food plants, such as the potato and tomato. Red peppers, chillis, and tobacco also belong to this family. Many species are poisonous, including Henbane, Thorn-apple and the nightshades. The flowers often have a central cone of yellow anthers. The fruits are usually berries.

BITTERSWEET
Solanum dulcamara

The small, purple flowers, each with a cone of yellow anthers, are unmistakable. The plant is mildly poisonous and climbs 30-200 cm high in hedges, woods or waste places. It is also called Woody Nightshade.
Flower: purple, with central cone of yellow anthers, 1-1·5 cm across; petals 5, pointed.
Flower arrangement: cyme.
Flowering time: June-September.
Leaf: ovate, pointed, stalked, with or without 1-4 lobes at base.
Fruit: berry, poisonous, shiny, ovoid, green when unripe, becoming yellow, then red.

BLACK NIGHTSHADE
Solanum nigrum

This annual is poisonous. The plant is hairless or with short hairs and grows up to 60 cm high on waste and cultivated land.
Flower: white, with central cone of yellow anthers, about 5 mm across; petals 5, pointed.
Flower arrangement: cyme.
Flowering time: July-September.
Leaf: ovate or diamond-shaped, pointed, stalked, toothed or not.
Fruit: berry, poisonous, black.

DEADLY NIGHTSHADE
Atropa bella-donna

This uncommon, highly poisonous, narcotic plant grows in dry places on chalk or limestone. It is stout, branched and grows up to 150 cm high.
Flower: lurid violet or greenish, 2·5-3 cm, nodding, bell-shaped, lobes pointed; sepals pointed.
Flower arrangement: solitary.
Flowering time: June-August.
Leaf: alternate, ovate, up to 20 cm.
Fruit: berry, 1·5-2 cm across, shiny, black, surrounded by persistent calyx.
Uses: the poisonous and medicinal properties of this plant have long been known. An infusion of plant juice was formerly dropped in women's eyes causing dilation of the pupils to produce a 'wide-eyed' look, hence the name *bella-donna* (beautiful lady).

FOXGLOVE FAMILY Scrophulariaceae

This large family contains about 3,000 species, most of which are native to the northern temperate regions. All the European species are herbaceous plants. Some species are semi-parasites, mainly attacking grasses. The flowers are usually tubular and irregular, with two lips, and they often attract specialist insect pollinators. The fruits are many-seeded capsules and the seeds are often extremely small. The mint family (pages 99-105) has some similar flowers, but the stems are square in section and the fruits are quite different. Many members of the foxglove family are cultivated in gardens. The snapdragons are good examples.

FOXGLOVE
Digitalis purpurea

The upright, softly hairy Foxglove plants with their spikes of purple bells are easy to recognize. They prefer acid soils and grow in open woodland, heaths and field borders. Children poke their fingers in the flowers to make a wish.
Flower: pinkish purple, sometimes white, 4-5 cm long, bell-shaped, shallowly 5-lobed, spotted and whiskery within; sepals 5.
Flower arrangement: long spike.
Flowering time: June-September.
Leaf: ovate, blunt-toothed, softly hairy above, woolly beneath; narrow border on each side of leaf-stalk.
Fruit: capsule, ovoid, with long, persistent style.
Uses: Foxgloves are still cultivated for the heart drug digitalin.

EYEBRIGHT
Euphrasia officinalis

The eyebrights are extremely difficult to tell apart. The following description covers a group of species. All are semi-parasites on grasses and grow 1-40 cm high in grassy places. The stems are hairy and branched. The plants are annuals.
Flower: white or mauve, with yellow and purple marks; 2-lipped, lower lip of 3 notched lobes; calyx 4-toothed; bracts leaf-like.
Flower arrangement: in leaf-axils.
Flowering time: May-September.
Leaf: circular or oval, deeply toothed.
Fruit: capsule, fringed with hairs.
Uses: this little plant was used in the past to treat eye disorders.

YELLOW RATTLE
Rhinanthus minor

This annual is semi-parasitic on grasses, amongst which it grows. The stem may be branched, 10-50 cm high and marked with black.
Flower: yellow, 1-1·5 cm long, tubular, 2-lipped, teeth of upper lip purple; calyx 4-toothed, inflated.
Flower arrangement: crowded with leaf-like bracts at top of stems.
Flowering time: May-August.
Leaf: oblong to lanceolate, toothed, opposite, stalkless, rough to touch.
Fruit: formed from enlarged calyx in which ripe seeds rattle.

GERMANDER SPEEDWELL
Veronica chamaedrys

Common in grassy places, hedges and woods, this small speedwell grows 7-25 cm high. The stem, with its 2 lines of hairs, lies along the ground, rooting at intervals and turning up at the tip. Speedwell flowers characteristically drop from the calyx at the lightest touch.
Flower: bright blue, white circle at centre, 1 cm across; petals 4, upper largest; sepals 4; bracts shorter than or equalling individual flower stalks; stamens 2.
Flower arrangement: raceme springing from leaf-axil.
Flowering time: March-July.
Leaf: oval, blunt-toothed, hairy, shortly stalked or stalkless.
Fruit: capsule, heart-shaped, hairy, shorter than sepals.

COMMON TOADFLAX
Linaria vulgaris

The flowers look like miniature yellow garden Snapdragons, to which Toadflax is related. Nectar is stored in the long spur behind the flower, accessible only to long-tongued bees. The plant is almost hairless, 30-80 cm high and grows along hedges and in grassy places.
Flower: yellow, top part of lower lip orange; 1·5-2·5 cm long; 2-lipped; spur almost straight; sepals 5, pointed.
Flower arrangement: dense raceme.
Flowering time: June-October.
Leaf: narrow, strap-shaped, pointed, alternate, 3-8 cm long, grey-green beneath.
Fruit: capsule, ovoid; seeds winged.
Uses: the plant is said to make a good fly poison if boiled in milk, the milk serving as an attractant.

GREAT MULLEIN
Verbascum thapsus

The leaves and stem of this tall biennial are thickly covered in white wool. It grows in dry, waste places and chalk grassland.
Flower: yellow, 1·5-3 cm across; petals 5, rounded, joined at base; sepals 5, pointed; 3 stamens covered with pale hairs, 2 hairless.
Flower arrangement: tall, dense spike.
Flowering time: June-August.
Leaf: oblong, pointed or rounded at tip; base running down stem as narrow border; basal leaves in rosette.
Fruit: capsule, enclosed by sepals; seeds tiny.

RED BARTSIA
Odontites verna

This is a semi-parasitic annual, much branched and downy. It grows to 50 cm in grassland, on waste land, and on the margins of cultivated fields. It obtains water and minerals from the roots of grasses.
Flower: purple-pink, about 1 cm, tubular; 2-lipped, lower lip 3-lobed; calyx of 4 teeth; bracts leaf-like.
Flower arrangement: in spikes, flowers pointing one way.
Flowering time: June-August.
Leaf: lanceolate, toothed, opposite, stalkless. Often tinged with maroon.
Fruit: capsule, seeds ridged.

BROOKLIME
Veronica beccabunga

Both stem and leaves of this water-loving plant are hairless and rather fleshy. The bases of the hollow stems creep and root in wet mud or shallow fresh water. The plant reaches a height of about 60 cm.
Flower: blue, 5-8 mm across; petals 4, upper petal larger than lower 3; sepals 4; stamens 2.
Flower arrangement: racemes in leaf-axils.
Flowering time: May-September.
Leaf: circular to oval, opposite, slightly fleshy, shiny, blunt-toothed, short-stalked.
Fruit: capsule, spherical, slightly notched.
Uses: the sharp-tasting leaves have been eaten as a substitute for Water-cress.

COMMON FIGWORT
Scrophularia nodosa

This is a rather tall, hairless plant growing in wet woods and ditches. The stem is 4-angled. The dull maroon flowers smelling of decay attract wasps by mimicking carrion. Normally 40-80 cm high, the plant may reach a height of about 1 m.
Flower: upper part maroon, lower green, 1 cm long; 5-lobed, upper 2 slightly longer; sepals 5, rounded.
Flower arrangement: branched clusters with sticky hairs on the flower stalks.
Flowering time: June-September.
Leaf: ovate, pointed, toothed, opposite, 6-13 cm long.
Fruit: capsule, ovoid, pointed.
Uses: the knobs on the rhizomes were taken as signs that this plant cured both piles and goitre.

IVY-LEAVED TOADFLAX
Cymbalaria muralis

This attractive little plant is a native of the Mediterranean region, but now established throughout Europe. It is a hairless trailing plant that grows best on dry rocks and walls.
Flower: 8-10 mm across; lilac with a white swelling (the palate) on lower lip and a yellow spot at the mouth. Borne singly on slender curving stalks. May-September.
Leaf: shiny green above, often purplish below: shaped like a tiny ivy leaf, about 2·5 cm across.

BROOMRAPE FAMILY Orobanchaceae

Members of this family are total parasites, without chlorophyll and drawing all their food supplies from the roots of various host plants. Stems and flowers are usually brown, yellow, or pinkish. Flowers are tubular and 2-lipped. Fruit is a capsule with numerous tiny seeds.

COMMON BROOMRAPE
Orobanche minor

One of several rather similar plants, which are best distinguished by their host plants and by the colour of their stigma lobes, this spiky annual is often abundant on roadside verges and other rough grassland. It parasitizes clovers and other legumes and also various composites. The upright pinkish-brown stem reaches 50 cm.
Flower: up to 18 mm long, yellowish brown with purple veins and spots: tube strongly and smoothly curved: stigma lobes purple. Clustered into a dense spike (sometimes just a few flowers). May-July, or even later.
Leaf: simple brown scales on stem.

BUTTERWORT FAMILY Lentibulariaceae

The members of this small family are insectivorous plants of water and wet places. The flowers are 2-lipped, with a prominent spur. Apart from the butterworts, the family contains the bladderworts – rootless water plants which have suction traps to catch small animals.

COMMON BUTTERWORT
Pinguicula vulgaris

This is a plant of wet heaths and moors and damp rocky places in general. It grows mainly in upland regions.
Flower: up to 12 mm wide: violet coloured and often mistaken for a violet when seen without the leaves; a white patch in the throat, and lower lip split into 3 lobes. Borne singly on slender, leafless stems up to 15 cm. May-July.
Leaf: pale yellowish green; elongated oval to 8 cm long. Margins inrolled. Forming a neat, flat rosette. Leaf surface covered with minute glands exuding a sticky fluid which traps insects and digests them. Leaf rosette dies in autumn and plant passes winter as a rootless bud just under the ground.

MINT FAMILY Labiatae

This large family contains about 3,500 species, concentrated in the Mediterranean region. Many members are aromatic and they include many popular kitchen herbs, such as Sage, Marjoram, and Thyme, as well as the mints. The stems are square in cross section and the leaves are in opposite pairs. The flowers are borne in whorls. The calyx is funnel-shaped and the petals are joined into an irregular, 2-lipped tube. The ovary is 4-lobed, each lobe producing a 1-seeded nutlet deep in the persistent calyx. Chemicals called terpenes are found in the family and they often suppress the growth of surrounding plants.

WATER MINT
Mentha aquatica

The often reddish stems of Water Mint are erect, rising from a creeping underground stem, and 15-90 cm tall. Water Mint grows in wet places.
Flower: mauve; petals 4; calyx 5-toothed, hairy; stamens long with red anthers.
Flower arrangement: shoots ending in dense, rounded heads.
Flowering time: July-October.
Leaf: ovate, teeth blunt or pointed, opposite, more or less hairy, stalked; if growing in water, the submerged leaves are rounded, toothless.
Fruit: 4 nutlets, smooth.
Uses: as with many other mints, Water Mint has been used as a stomach medicine and as a herb for strewing on floors of houses.

GROUND IVY
Glechoma hederacea

Many species in the mint family have purplish, 2-lipped flowers, all looking rather similar. Ground Ivy is distinguished by having only 2-5 flowers in each whorl and the top petal-lobe being flat, not hooded. Creeping and rooting, it grows in woods and waste places.
Flower: pale violet, 1·5-2 cm long; 2-lipped, upper lip flat, not hooded, lower purple-spotted; calyx 5-toothed.
Flower arrangement: in loose whorls, directed to one side.
Flowering time: March-May.
Leaf: kidney-shaped to almost circular, blunt-toothed, often tinged purple, opposite, long-stalked.
Fruit: 4 nutlets, smooth.

MARJORAM
Origanum vulgare

This is a plant of dry places on chalk or limestone, smelling aromatic when crushed, due to tiny translucent oil-filled glands on the leaves, seen if held to the light. The 4-angled stems, 30-80 cm long, are erect and branched in the upper parts.
Flower: mauve, 6-8 mm across; 4-lobed; calyx 5-toothed; stamens 4; bracts often purple.
Flower arrangement: in flat-topped or rounded clusters.
Flowering time: July-September.
Leaf: ovate, opposite, slightly toothed or not, short-stalked.
Fruit: 4 nutlets, each ovoid.
Uses: this native species is more pungent than the cultivated, frost-tender species from the Mediterranean used in cooking. Tea can be made from the dried leaves.

MEADOW CLARY
Salvia pratensis

This hairy and slightly aromatic plant, a relative of the cultivated Sage, grows in rough, calcareous grassland. Confined to the most westerly parts of Europe, from Denmark southwards, it is most abundant in France. It reaches a height of 1 m.
Flower: brilliant blue, up to 2·5 cm long, with a strongly curved upper lip forming a distinct hood: stamens and style project beyond petals. Borne in whorls on stiff, leafless stems. June-July. Some plants carry only female flowers.
Leaf: basal leaves up to 15 cm long, in a rosette: oval to lanceolate and very wrinkled. Stem leaves smaller, the upper ones stalkless.
Fruit: 4 smooth triangular nutlets. Wild Clary (*S. horminoides*) of France and southern Britain is similar, but flowers are more purplish-blue with shorter style: leaves are more jagged.

SELFHEAL
Prunella vulgaris

Selfheal is a low-growing, slightly hairy plant of grassland, often tinged purple. The stem is 4-angled, creeping in the lower part and ending in an oblong head of flowers. The plant has no smell when crushed.
Flower: violet, 1-1·5 cm long; 2-lipped, upper lip hooded, lower 3-lobed; calyx of 5 teeth in 2 lips; bract beneath each flower almost circular, hairy and purplish.
Flower arrangement: in oblong or squat head at top of stem, pair of leaves immediately beneath.
Flowering time: June-September.
Leaf: ovate, tip pointed, 2-5 cm long, shallowly toothed or not, opposite, stalked.
Fruit: 4 nutlets, each oblong.

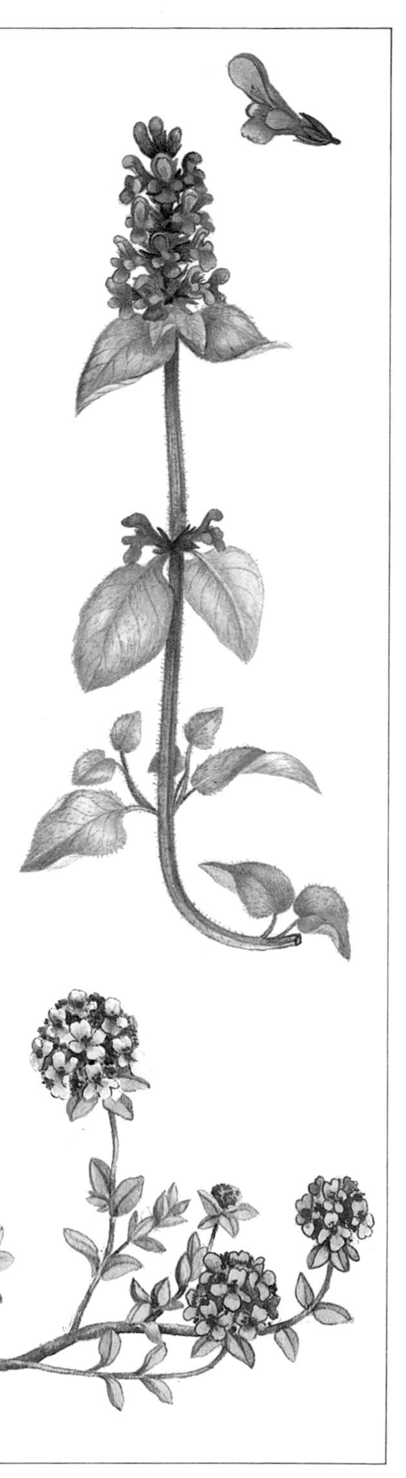

WILD THYME
Thymus drucei

A small relative of the true Thyme (*Thymus vulgaris*) which is used in cooking, this creeping plant has wiry stems and forms thick carpets in dry grassland – where it very commonly covers ant hills – and on dunes and rocks. It rarely exceeds 7 cm in height. The species may be separated from several other very similar plants by looking at the upright shoots: they are very hairy on 2 opposite sides and almost hairless on the other 2 sides. Large Thyme (*T. pulegioides*) is hairy only on the angles of the stem, while Breckland Thyme (*T. serpyllum*) is hairy all round the stem.
Flower: pink or purple; 3-5 mm across, with stamens protruding well beyond petal tube. In small whorls at the top of upright shoots. May-August.
Leaf: oval, 4-8 mm long: dark shining green and sometimes hairy.
Fruit: smooth ovoid nutlets.

LAVENDER
Lavandula angustifolia

This is a low-growing, much branched shrub famed for its aromatic oils which are widely used in perfumes. It is a native of southern Europe, where it grows on dry, rocky hillsides, usually on limestone. It reaches heights of about 75 cm. It is cultivated in large fields for the perfume industry, the strongly-scented flowers being gathered just as they begin to open. They are then distilled to yield lavender oil. The bushes are regularly trimmed into neat domes about 50 cm high to encourage fresh new shoots. Lavender is also a popular garden plant in many parts of Europe.

Flower: purplish-blue (lavender), 3-5 mm across: upper lip not hooded. Calyx tubular and grey-green. Borne in dense oblong spikes at the top of downy, grey-green stems. June-August.

Leaf: grey-green and downy, more or less linear: up to 4 cm long.

FRENCH LAVENDER
Lavandula stoechas

This much-branched, strongly-scented undershrub is a close relative of the Common Lavender, but often develops into larger and more straggly bushes. Another native of the Mediterranean region, it thrives on dry, stony hillsides. It is often a major associate of the cork oak trees on the acidic soils of south-east France. It reaches about 1 m. The flowers are sometimes gathered to provide oil for perfumes.

Flower: individual flowers are similar to those of the Common Lavender, grouped into dense oblong spikes, but the top of each spike carries a number of large purple bracts which help to attract insects to the spikes. Smaller bracts among the flowers are papery and purplish-green. April-July.

Leaf: greyish-green and velvety: more or less linear to about 4 cm.

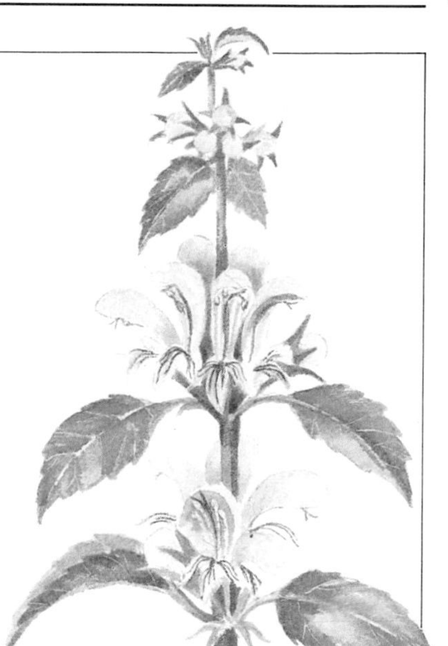

YELLOW ARCHANGEL
Galeobdolon luteum

This rather stiff and hairy plant, a close relative of the White Deadnettle, is found in woodlands and shady hedge-banks on rich, damp soils, especially in calcareous regions. It reaches a height of about 60 cm. Long creeping runners spread out and take root in all directions, producing extensive patches of the plant, often with little else growing amongst it.
Flower: about 2 cm long, with upper lip forming a distinct hood: bright yellow with rust-coloured streaks on the 3-lobed lower lip. Borne in open whorls, well separated on upper parts of stem.
Leaf: 4-7 cm long, oval and coarsely toothed.
Fruit: 4 triangular nutlets, concealed, as in all labiates, deep in the persistent calyx.

HEDGE WOUNDWORT
Stachys sylvatica

The tough stems of this rather tall plant are roughly hairy and 4-angled. The plant grows 30-120 cm high in the shadier parts of hedgerows, and also in woods and waste places. If crushed the whole plant, in particular the creeping, underground stem, gives off a foul smell.
Flower: light maroon with white marks, 1-1·5 cm; 2-lipped; calyx with 5 narrow teeth and glandular hairs.
Flower arrangement: spike of whorls usually with about 6 flowers in each whorl; upper whorls closely spaced.
Flowering time: July-August.
Leaf: ovate, tip pointed, toothed, stalked, opposite, sparsely hairy.
Fruit: 4 nutlets.
Uses: the common name refers to the old use of this plant in staunching bleeding wounds. The plant also contains antiseptic properties.

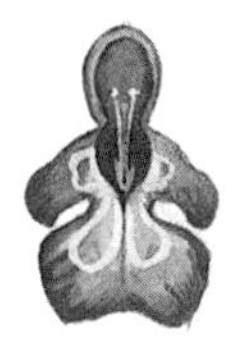

RED DEAD-NETTLE
Lamium purpureum

Red Dead-nettle is a softly hairy weed of cultivated ground. It has a 4-angled stem and grows 10-45 cm high. It is an annual, branching freely from the base. Dead-nettles, as the name implies, have no stinging hairs.
Flower: mauve-purple, 1-1·5 cm long; 2-lipped, upper hooded, lower spotted; ring of hairs near flower base; calyx 5-toothed.
Flower arrangement: in dense whorls with leaf-like bracts.
Flowering time: March-October.
Leaf: ovate to almost circular, blunt-toothed, often tinged purple, surface puckered, opposite, stalked.
Fruit: 4 nutlets, 3-angled.

WHITE DEAD-NETTLE
Lamium album

Although a very common plant of waste places and roadsides, the tight whorls of plump-looking, pure white flowers make this an attractive plant. The 4-angled stem is unbranched, hairy and 20-80 cm high. Unlike the Red Dead-nettle, this species is a perennial plant and has a mat of creeping rhizomes which spread through the soil and throw up new shoots to produce dense clumps.
This dead-nettle is not related to the Common Nettle and does not sting, although the leaves are similar.
Flower: white, 2-2·5 cm long; 2-lipped, upper lip pronouncedly hooded, hairy, lower lip with 2 or 3 short teeth each side; calyx with 5 narrow, pointed teeth.
Flower arrangement: in compact whorls.
Flowering time: May-December.
Leaf: ovate, tapering to point, coarsely toothed, opposite, stalked.
Fruit: 4 nutlets, 3-angled.
Uses: country children have made whistles out of the hollow stems and they also enjoy sucking nectar from the flowers.
A closely related species (*L. maculatum*), whose leaf blades commonly bear a large white blotch, is often cultivated in gardens. It has pinkish flowers.

COMMON HEMP-NETTLE
Galeopsis tetrahit

The branched, roughly hairy stem has red or yellow glandular hairs and is swollen where the leaf stalks join. The swellings contain special cells to enable the plant to make slight movements. This plant is 10-100 cm tall and found on cultivated land, less often in damp places. It is an annual.
Flower: mauve with darker markings, occasionally white or pale yellow, 1·5-2 cm long; 2-lipped, upper lip hooded, lower 3-lobed, middle lobe notched or not; calyx with 5 narrow teeth.
Flower arrangement: in whorls.
Flowering time: July-September.
Leaf: ovate, tapering to point, blunt-toothed, stalked, blue-green.
Fruit: 4 nutlets, 3-angled.

PLANTAIN FAMILY Plantaginaceae

This is a small family with about 300 species, mainly from temperate areas or tropical mountains. The small flowers are clustered tightly into spikes and are mainly wind-pollinated. They produce large amounts of powdery pollen and, together with the grasses, the plantains are major causes of hay-fever. Among the plantains the lowest flowers on the spike open first and they die before the uppermost flowers have opened. A ring of pendulous stamens thus moves slowly up the spike. Several species of plantain are stubborn weeds, although Ribwort Plantain is quite nutritious and a useful component of pasture land.

HOARY PLANTAIN
Plantago media

Growing in fairly dry grassy places, especially on chalk and limestone soils, this species resembles the Greater Plantain, but has markedly furry, greyish-green leaves.
Flower: white and scented, with lilac or white anthers, producing a distinctly pale spike. The latter is 2-8 cm long and borne on a smooth stalk to 30 cm. Pollinated partly by insects.
Flowering time: May-August.
Leaf: greyish-green and clothed with fine greyish hairs. Ovate to elliptical, usually 4-6 cm long, gradually narrowing into a short stalk. In a rosette and usually pressed close to the ground.
Fruit: a 4-seeded capsule.

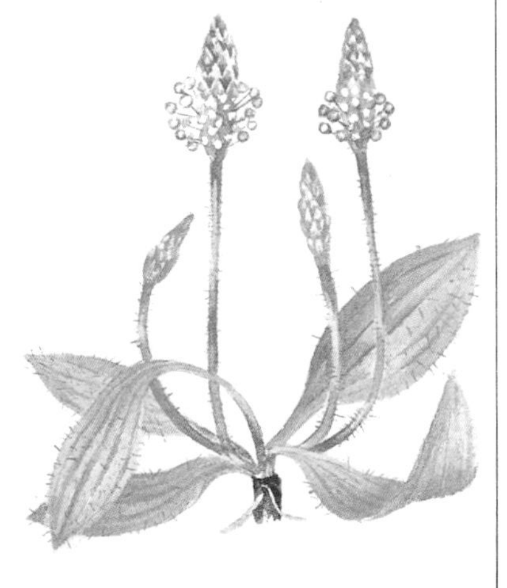

GREATER PLANTAIN
Plantago major

The rosettes of tough, ribbed leaves, often flattened by, but very resistant to trampling, are seen at the sides of almost every path and piece of trodden ground.
Flower: greenish yellow, tiny; anthers mauve, becoming yellow, protruding, wind-pollinated.
Flower arrangement: in long, dense, stalked spike, 10-15 cm, encircled by whorl of anthers.
Flowering time: May-September.
Leaf: broadly elliptic, in rosette; 10-15 cm long, usually without teeth, strongly ribbed, abruptly narrowed into broad stalk.
Fruit: capsule, 2-5 mm.
Uses: when in seed the flower spikes may be hung up for cage-birds to feed on.

RIBWORT PLANTAIN
Plantago lanceolata

The ribbed leaves of this plantain are long and narrow, forming rosettes. The plant grows in grassy places.
Flower: browny-black, tiny; anthers cream, protruding, wind-pollinated.
Flower arrangement: in dense, short spike of 1-2 cm, on grooved, tough stalk, much longer than leaves.
Flowering time: April-August.
Leaf: lanceolate, in rosette; 10-15 cm long, usually without teeth, strongly ribbed. There is no obvious demarcation between blade and stalk.
Fruit: capsule, about 5 mm.
Uses: the seeds become slimy-coated when wet. In France the coating was used as a fabric stiffener, especially of muslin.

HONEYSUCKLE FAMILY Caprifoliaceae

Most of the species in this family are shrubs, some of them climbing. The flowers are often tubular, and the fruits are usually berries.

HONEYSUCKLE
Lonicera periclymenum

A climbing shrub of up to 6 m, Honeysuckle may also be found sprawling over the ground. It grows in hedges and woods and is planted in gardens for the scented flowers.
Flower: cream, tinged with pinky red; trumpet-shaped with 4 upper lobes, 1 lower; 4-5 cm long, stamens and style protruding.
Flower arrangement: in head, flowers directed outwards.
Flowering time: June-October.
Leaf: elliptic or oblong, tip usually pointed, stalkless, 3-9 cm long, dark green above, paler beneath.
Fruit: tight head of red berries.

BELLFLOWER FAMILY Campanulaceae

Many of the flowers in this family are blue and bell-shaped and are grown as garden ornamentals. A sugar (*inulin*) occurs in the sap that is identical to that found in plants of the Daisy family. The two families are thought to be closely related.

CLUSTERED BELLFLOWER
Campanula glomerata

This erect, downy plant grows on rough calcareous grasslands and reaches heights of about 30 cm.
Flower: bell-shaped, with mouth upwards: 15-20 mm long with 5 pointed lobes. Deep purplish-blue. Borne in dense clusters at top of normally unbranched stem.
Leaf: basal leaves oval to triangular; long stalked: blade to 8 cm long and lightly toothed. Stem leaves ovate to lanceolate, upper ones stalkless.

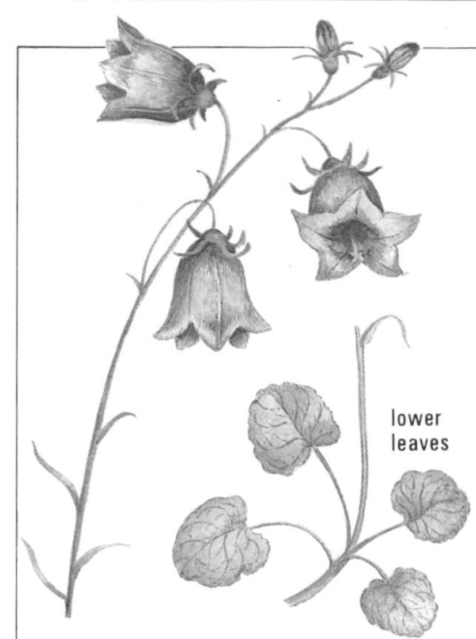

HAREBELL
Campanula rotundifolia

Called Bluebell in Scotland, the Harebell is unrelated to the English Bluebell (page 125). It is a very slender, hairless plant of dry, grassy places and grows 15–40 cm.
Flower: pale blue, 1-2 cm, bell-shaped with 5 pointed lobes, nodding; calyx with 5 narrow, pointed teeth; stigmas 3.
Flower arrangement: solitary or in loose cluster, on thread-like stalk.
Flowering time: July-September.
Leaf: very narrow, pointed, unstalked on stem; basal leaves ovate or circular, blunt-toothed, stalked.
Fruit: capsule, papery, 3-5 mm.

BEDSTRAW FAMILY Rubiaceae

This large family contains many tropical trees and shrubs, including the coffee plant. European members are all herbs. Leaves appear to be whorled, but each 'whorl' has just two opposite leaves and a number of leaf-like stipules. The flowers are always small, with 4 or 5 joined petals.

CLEAVERS
Galium aparine

This scrambling plant of 15-180 cm clings to other vegetation, often in dense masses, with minute hooked bristles, rough and tacky to the touch. Also called Goose-grass, it grows in hedges and woods and is a weed of cultivation.
Flower: white, 2 mm across; petals 4.
Flower arrangement: in clusters of 2-5, on long stalks, with whorl of leaf-like bracts.
Flowering time: June-August.
Leaf: in whorls of 6-9, narrow, tip ending in bristle, single-veined.
Fruit: of 2 joined lobes, 3-6 mm, covered with tiny, white, hooked bristles; purplish when ripe.

stem section, showing hooks

fruit

LADY'S BEDSTRAW
Galium verum

The stems of this yellow-flowered bedstraw are more or less erect, 15-100 cm high, and without the minute, hooked bristles used by other bedstraw species for climbing. The plant grows in dry grassland.
Flower: golden yellow, 2-4 mm across; petals 4.
Flower arrangement: in many-flowered clusters at ends of branches.
Flowering time: July-August.
Leaf: in whorls of 8-12, grass-like, 6-25 mm long, bristle-tipped; margins rolled under; dark green and shiny above, paler beneath.
Fruit: of 2 joined lobes, 1·5 mm, hairless, black when ripe.
Uses: this bedstraw in particular was used to turn and colour the milk in cheese-making.

WOODRUFF
Galium odoratum

This fragrant woodland plant, smelling strongly of vanilla or fresh hay, grows on both calcareous soils and heavy clays, but avoids sandy areas. Its erect, unbranched stems spring from creeping rhizomes and reach 45 cm.
Flower: white and funnel-shaped, deeply 4-lobed, to 6 mm across. Packed into umbel-like heads. May-June.
Leaf: lanceolate and hairless but with spiny edges, bright green: 6-8 in a whorl.
Fruit: rounded, 2-3 mm: with hooks.

fruit

TEASEL FAMILY Dipsacaceae

This small family includes the various kinds of scabious as well as the teasels. The tight heads of flowers resemble those of the daisy family, except for the stamens which stand out from each tubular flower. Each flower head has a cluster of bracts under it, and there may be small bracts among the flowers as well. In the Teasel itself, these small bracts are spiny.

FIELD SCABIOUS
Knautia arvensis

The beautiful mauve flower-heads, made up of many tiny flowers, are often larger than the other scabious species. The erect plants grow 25-100 cm high on dry banks and fields, and the stem has rough bristles, at least at the base.

Flower: mauve, 4-lobed; outer flowers larger than inner; stamens protruding.

Flower arrangement: in flattish head, 3-4 cm across, on long stalk; bracts directly below head ovate, in 2 overlapping layers, not reaching edge of head.

Flowering time: July-September.

Leaf: variable, deeply lobed, opposite, hairy; basal leaves often forming rosette, some unlobed.

Fruit: 5-6 mm, crowned by 8 bristles; dispersed by ants, which find them attractive.

DEVIL'S BIT SCABIOUS
Succisa pratensis

This scabious has undivided leaves and an erect, slightly hairy stem 15-100 cm high. It grows in damp places, including woods. The curious root gives the plant its name. After the first year's growth the tip of the thick root falls away leaving an abrupt end, as if bitten off.

Flower: mauve, 4-lobed; outer flowers not much larger than inner; stamens protruding.

Flower arrangement: in domed head, 1·5-2·5 cm across, on long stalk; sepal-like bracts lanceolate, in 2 overlapping layers, reaching edge of head. Leafy bracts among flowers.

Flowering time: June-October.

Leaf: elliptic, tip pointed or rounded, opposite, untoothed or with a few teeth; basal leaves form rosette.

Fruit: 5 mm, crowned by 5 bristles.

DAISY FAMILY Compositae

This enormous family of about 25,000 species includes many common weeds as well as numerous cultivated plants. The family is characterized by tiny flowers (florets) which are packed into tight heads. The florets are of 2 kinds – a simple tube (disc floret) or a tube with a strap-like outgrowth on one side (ray floret). The heads may have only disc florets (thistle), only ray florets (dandelion), or a mixture (daisy). Florets may be male, female, or hermaphrodite. With sepal-like bracts underneath, each head may look like a single flower. The fruits (achenes) often have a parachute of hairs (pappus) for wind dispersal. Cultivated species include lettuces, asters, and sunflowers.

DAISY
Bellis perennis

Familiar in the short grass of meadow, lawn and verge, the Daisy flower stalks rise straight from a rosette of leaves. The flowers close at night and in wet weather.
Flower: ray florets white, tinged pink beneath, many; disc florets yellow; head 1·5-3 cm across.
Flower arrangement: head solitary, on long stalk.
Flowering time: March-October.
Leaf: spoon-shaped.
Fruit: achene, 1·5-2 mm, downy.

PINEAPPLEWEED
Matricaria matricarioides

Named for the strong aromatic smell, supposed to resemble pineapple scent, when crushed, this low-growing annual is a weed of cultivation and waste places, especially where trampling is heavy. Introduced to Europe, probably from north-east Asia, it has ferny leaves.
Flower: yellow-green; no ray florets: heads domed, 5-8 mm across and resembling a miniature pineapple: sepal-like bracts rounded and pale-bordered.
Flower arrangement: heads loosely clustered on branched stems.
Flowering time: June-July.
Leaf: finely dissected, fern-like.
Fruit: a small achene.

SCENTLESS MAYWEED
Tripleurospermum maritimum

This mayweed has daisy-like flower-heads and ferny leaves. It has no smell when crushed. The branched stem is 10-30 cm high, usually prostrate. The whole plant is hairless Grows on waste ground and as an arable weed.
Flower: ray florets white; disc florets yellow; heads 1·5-5 cm across; sepal-like bracts with narrow, papery margins.
Flower arrangement: heads loosely on branched stems.
Flowering time: July-September.
Leaf: ferny, very finely dissected.
Fruit: achene, with 2 dark dots near top.

COMMON RAGWORT
Senecio jacobaea

Neglected fields are often overgrown with this biennial species. It is also found on waste ground and sand dunes. The grooved stem branches above the middle and grows 30-150 cm high.
Flower: ray florets yellow, 12-15; disc florets orange; heads 1·5-2·5 cm across; sepal-like bracts black-tipped.
Flower arrangement: heads in flat-topped corymbs.
Flowering time: June-October.
Leaf: deeply lobed, lobes blunt-toothed, end lobe rounded; dark green, hairless or with sparse hairs beneath.
Fruit: achene, ribbed, with pappus.
The Oxford Ragwort (*S. squalidus*) is a shorter and bushier annual, with more slender, pointed leaf lobes. Bracts with conspicuous black tips. A native of southern Italy, it is now well established on waste ground in many parts of Europe: particularly common on old walls and along railway lines. It flowers May-December.

OX-EYE DAISY
Leucanthemum vulgare

The Ox-eye Daisy has large, daisy-like flower-heads, smaller in exposed places. The almost hairless stems grow 20-70 cm tall in grassy places.
Flower: ray florets white; disc florets yellow; head 2·5-5 cm across; sepal-like bracts with purplish borders.
Flower arrangement: heads solitary, on long stalks.
Flowering time: June-August.
Leaf: Often forming non-flowering rosettes, dark green, toothed; lower stem leaves rounded or spoon-shaped, long-stalked; upper leaves oblong, stalkless.
Fruit: achene, pale grey, ribbed.

flower bud

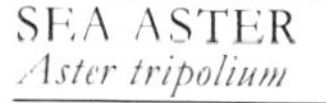

SEA ASTER
Aster tripolium

Closely related to the Michaelmas Daisy of gardens and with similar flowers, Sea Aster has fleshy leaves and grows 15-100 cm high in salt marsh and on sea cliffs.
Flower: ray florets mauve or whitish, many or none; disc florets yellow; head 8-20 mm across; sepal-like bracts rounded at tip.
Flower arrangement: in loose corymbs.
Flowering time: July-October.
Leaf: fleshy, hairless; upper leaves narrow, pointed, stalkless; lower leaves rounded at tip, stalked.
Fruit: achene, with brownish pappus.
The cultivated Michaelmas Daisies originated in North America. They often escape from gardens and become established in waste places, especially along railway lines and around rubbish dumps.

flower head without ray florets

GROUNDSEL
Senecio vulgaris

Groundsel grows on waste and cultivated ground. The branched stem is 8-45 cm high. The common name is very old, coming from an Anglo-Saxon word meaning 'ground swallower', from the way this weed spreads. It is an annual.

Flower: yellow; disc florets only, occasionally rayed; heads about 4 mm across; sepal-like bracts dark-tipped, outer short.

Flower arrangement: heads in loose clusters.

Flowering time: all year.

Leaf: lobed, the lobes irregularly toothed; upper leaf bases clasping stem; hairless or slightly hairy.

Fruit: achene, with pappus forming clocks.

CREEPING THISTLE
Cirsium arvense

Creeping Thistle is a stubborn weed on cultivated and waste land. The creeping roots send up new plants and produce dense clumps of upright stems. Ploughing does not destroy the weed, for cutting through the roots simply increases the number of pieces that can throw up new shoots.

The grooved stem is not continuously spiny-ridged, as in the common Spear Thistle, and grows 30-120 cm high.

Flower: mauve; disc florets only; head 1·5-2·5 cm across; sepal-like bracts sharply pointed but not spiny, purple-tinged. Fragrant. Male and female florets are sometimes borne on separate plants, the male plants then never producing the familiar thistledown.

Flower arrangement: solitary, or in clusters.

Flowering time: July-September.

Leaf: lanceolate, margin undulating, very spiny, hairless above, white-hairy or not beneath.

Fruit: achene; pappus long, fawn.

WELTED THISTLE
Carduus acanthoides

The common name refers to the matted white hairs on the stem and undersides of the leaves, though the degree of hairiness varies. Thin, spiny ridges run up the stem, ending short of the flower head. The plant grows 30-150 cm tall in hedges, and on verges and stream-sides, mainly on calcareous soils. It is a biennial.
Flower: reddish purple; disc florets only; head 2-2·5 cm across; sepal-like bracts many, sharply pointed.
Flower arrangement: in loose clusters of 3-5.
Flowering time: June-August.
Leaf: margins spiny, wavy, white hairs beneath; upper leaves narrow, lower deeply lobed.
Fruit: achene, with long pappus.

SPEAR THISTLE
Cirsium vulgare

Thin, spiny ridges run down the stems of this common biennial. The stems have white, woolly hairs and grow 30-150 cm tall in fields and waste places, and by roadsides.
Flower: mauve; disc florets only; head 3-5 cm across; sepal-like bracts spiny, hairy.
Flower arrangement: solitary, or cluster of 2-3.
Flowering time: July-October.
Leaf: lanceolate, margins wavy, very spiny; end segment sword-shaped; white-hairy or rough beneath.
Fruit: achene, yellow with black streaks; pappus long, white.

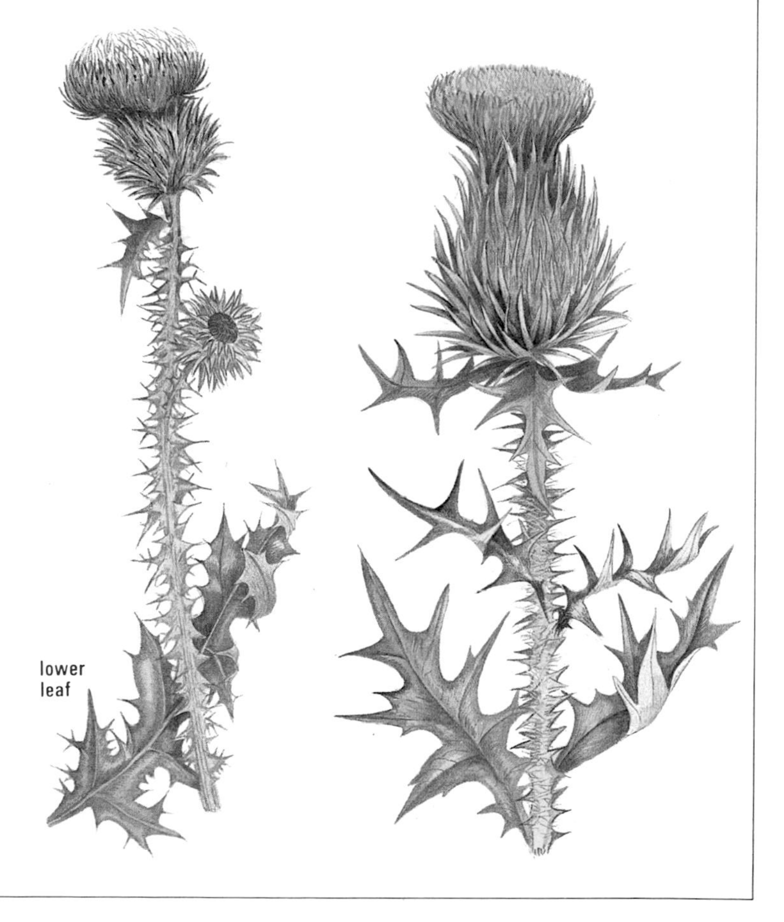

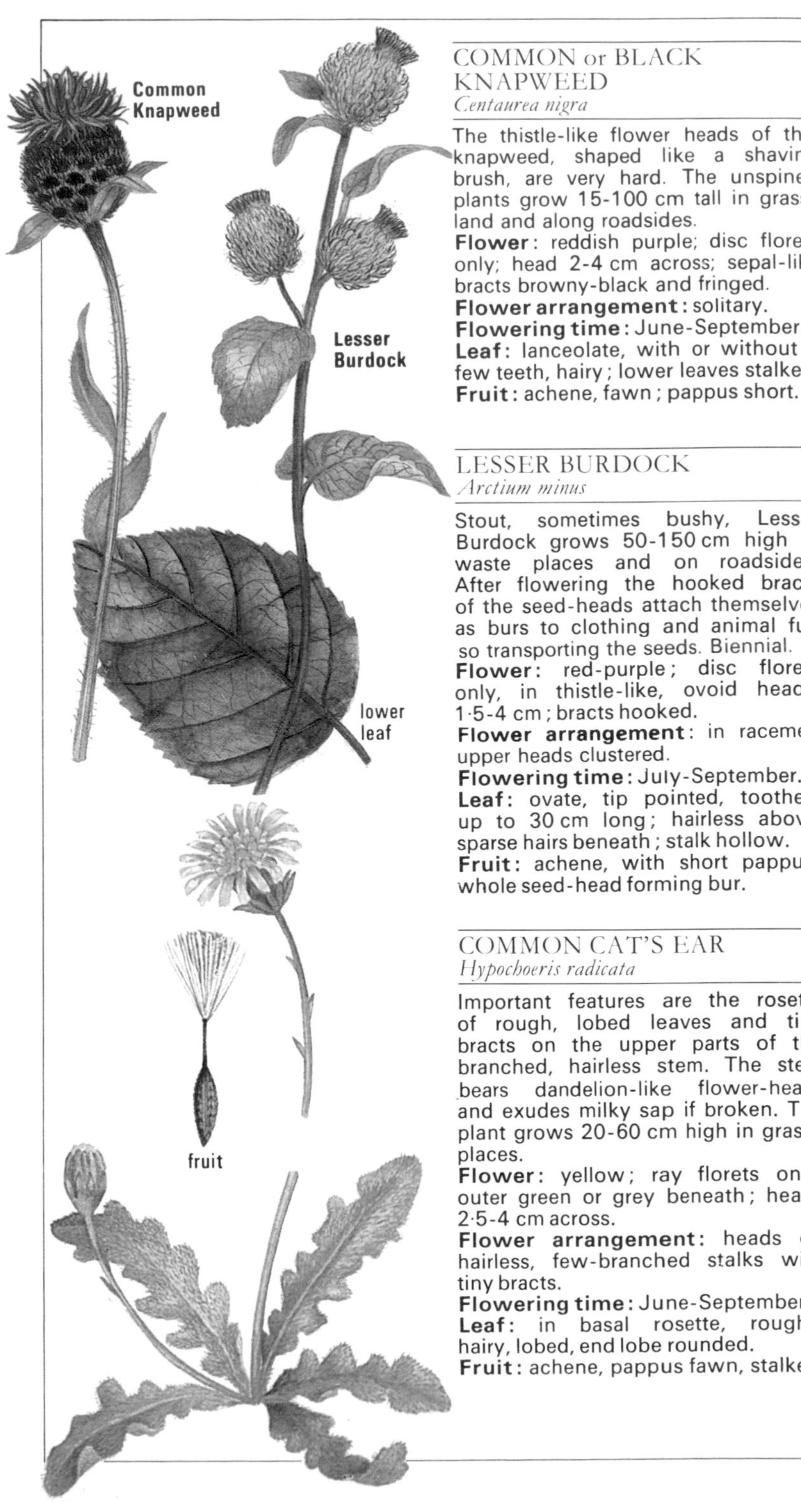

COMMON or BLACK KNAPWEED

Centaurea nigra

The thistle-like flower heads of this knapweed, shaped like a shaving brush, are very hard. The unspined plants grow 15-100 cm tall in grassland and along roadsides.

Flower: reddish purple; disc florets only; head 2-4 cm across; sepal-like bracts browny-black and fringed.

Flower arrangement: solitary.

Flowering time: June-September.

Leaf: lanceolate, with or without a few teeth, hairy; lower leaves stalked.

Fruit: achene, fawn; pappus short.

LESSER BURDOCK

Arctium minus

Stout, sometimes bushy, Lesser Burdock grows 50-150 cm high in waste places and on roadsides. After flowering the hooked bracts of the seed-heads attach themselves as burs to clothing and animal fur, so transporting the seeds. Biennial.

Flower: red-purple; disc florets only, in thistle-like, ovoid heads, 1·5-4 cm; bracts hooked.

Flower arrangement: in racemes; upper heads clustered.

Flowering time: July-September.

Leaf: ovate, tip pointed, toothed, up to 30 cm long; hairless above, sparse hairs beneath; stalk hollow.

Fruit: achene, with short pappus; whole seed-head forming bur.

COMMON CAT'S EAR

Hypochoeris radicata

Important features are the rosette of rough, lobed leaves and tiny bracts on the upper parts of the branched, hairless stem. The stem bears dandelion-like flower-heads and exudes milky sap if broken. The plant grows 20-60 cm high in grassy places.

Flower: yellow; ray florets only, outer green or grey beneath; heads 2·5-4 cm across.

Flower arrangement: heads on hairless, few-branched stalks with tiny bracts.

Flowering time: June-September.

Leaf: in basal rosette, roughly hairy, lobed, end lobe rounded.

Fruit: achene, pappus fawn, stalked.

DANDELION
Taraxacum officinale

There are many species of dandelion, all very variable and difficult to identify. This description covers a group of species. Dandelions grow in grassy and waste places. The pappus forms the familiar 'clocks' of children's games.
Flower: yellow; ray florets only, outer brown or mauve beneath; heads 3-7·5 cm across; outer sepal-like bracts turned down or spread out.
Flower arrangement: heads solitary, stalk unbranched, stout, hollow, exuding milky sap if broken.
Flowering time: March-October.
Leaf: in basal rosette, lobes deep, toothed, pointing down, hairless.
Fruit: achene, with stalked pappus.

PERENNIAL SOW-THISTLE
Sonchus arvensis

The plant has dandelion-like flower-heads, and leaves with rounded basal lobes. The hollow stem is 30-150 cm tall with sticky hairs on the upper parts. If cut, the stem oozes milky sap. The plant grows in waste places, by streams and on the coast.
Flower: golden yellow; ray florets only; heads 4–5 cm across; sepal-like bracts with sticky hairs.
Flower arrangement: heads in loose, flat-topped clusters.
Flowering time: July-October.
Leaf: oblong, lobed, margins weakly spiny; leaf bases clasping stem with 2 rounded lobes.
Fruit: achene, with white pappus.

SMOOTH SOW-THISTLE
Sonchus oleraceus

This annual has dandelion-like flower-heads and pointed basal lobes to the leaves. The stem is hollow, 20-150 cm high and usually lacks sticky hairs. If cut, it oozes milky sap. The plants grow in waste places and as a weed of cultivation.
Flower: pale yellow; ray florets only; heads 2-2·5 cm across; sepal-like bracts usually without sticky hairs.
Flower arrangement: heads clustered.
Flowering time: June-August.
Leaf: variable in shape, often deeply lobed, margins weakly spiny; leaf bases clasping stem with 2 pointed lobes; blue-green.
Fruit: achene, with white pappus.

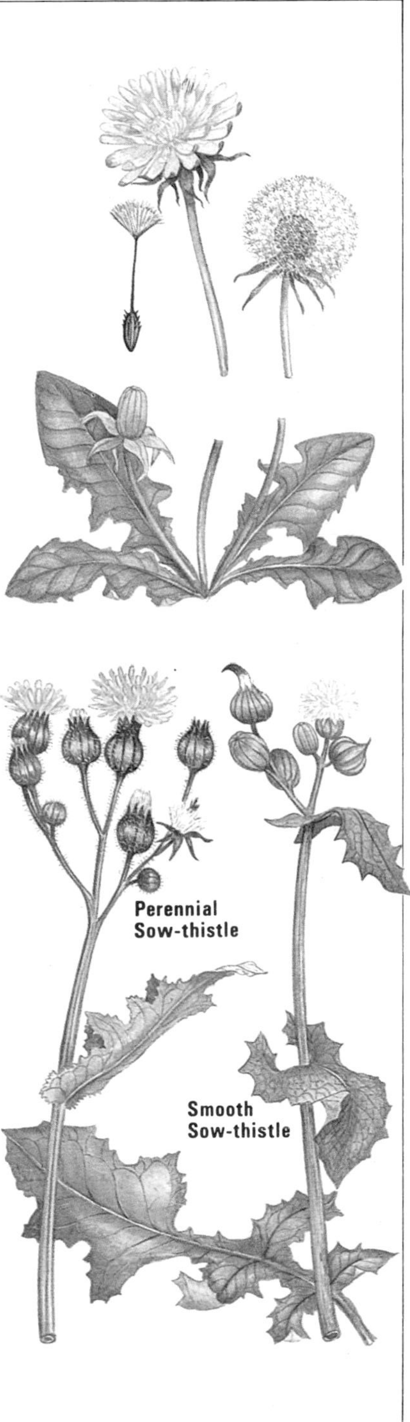

NIPPLEWORT
Lapsana communis

The flower-heads are like small dandelions on branched stalks, and the leaves grow up the stem. The plant can grow up to 90 cm high in shady, waste places. The common name is said to come from the shape of the flower buds. An annual.
Flower: yellow; ray florets only; heads 1·5-2 cm across; sepal-like bracts narrow, tips rounded, dark-striped down centre.
Flower arrangement: heads on branched stalks.
Flowering time: July-September.
Leaf: alternate, toothed; lower leaves with toothed lobes; all often slightly hairy.
Fruit: achene, ribbed, no pappus.

HAWKWEED
Hieracium Sect. vulgata

Hawkweed species are extremely numerous and difficult to identify. A section containing some of the most common species is described here. The branched, hairy stem grows 15-80 cm tall and exudes milky sap if broken. Hawkweeds are found in a wide variety of places.
Flower: yellow; ray florets only, Dandelion-like; sepal-like bracts dark green, glandular, hairy.
Flower arrangement: heads on branched, glandular hairy stalks.
Flowering time: June-September.
Leaf: elliptic to lanceolate, variably toothed, not lobed, both ends narrowed, hairy in parts; basal leaves in rosette.
Fruit: achene; pappus brownish.

AUTUMNAL HAWKBIT
Leontodon autumnalis

The outer petals of the dandelion-like flower-heads are often striped red beneath. The stems are branched, usually hairless, with tiny bracts towards the top. If broken, the stem exudes milky sap. The plants grow 5-60 cm high in grassy places.
Flower: yellow; ray florets only, striped red beneath; heads 1-3·5 cm across; buds erect; sepal-like bracts hairless (with dark hairs on mountain plants).
Flower arrangement: heads on branched, usually hairless stalks.
Flowering time: June-October.
Leaf: in rosette, lobed or with widely spaced teeth, tip pointed, hairless or slightly hairy.
Fruit: ribbed achene, with pappus.

SMOOTH HAWK'S-BEARD
Crepis capillaris

The branched stems of this rather slender, dandelion-like plant are 20-100 cm high. It grows in grassland and waste places.
Flower: bright yellow; ray florets only, outer often reddish beneath; heads 1-2·5 cm across; sepal-like bracts lanceolate, with black bristles or not, in 2 layers, outer bracts pressed close to inner.
Flower arrangement: loose clusters.
Flowering time: June-September.
Leaf: variable; lower leaves with many narrow lobes or toothed; upper leaves narrow, stalkless, with base having 2 pointed lobes clasping stem.
Fruit: achene, curved, 10-ribbed; pappus white.

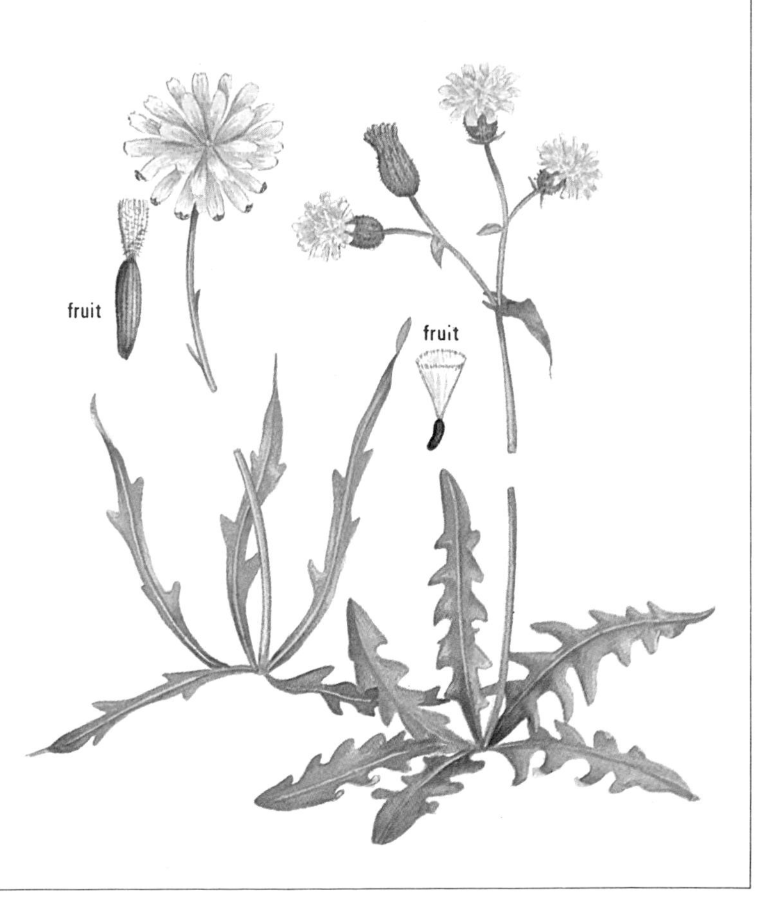

GOAT'S-BEARD
Tragopogon pratensis

The solitary dandelion-like flower-head of this biennial is overtopped by long bracts and closes around noon. The stem bears grass-like leaves and grows 30-70 cm high in grass- and wasteland.
Flower: yellow; ray florets only; heads broad; sepal-like bracts longer than head, narrow.
Flowering time: June-July.
Leaf: grass-like, often clasping stem at base, white-veined, hairless.
Fruit: achene, with stalked pappus, forming large clock.

COLTSFOOT
Tussilago farfara

Coltsfoot is found on waste and arable land, especially on clay soil. The flowering stems appear before the leaves in early spring, rising from a creeping, underground stem, as do the leaves later on.
Flower: yellow; disc florets surrounded by many narrow ray florets; head 1·5-3·5 cm across, drooping after flowering.
Flower arrangement: head solitary; stalk thick, scaled, 5-15 cm high.
Flowering time: February-April.
Leaf: appearing after flowers, 10-30 cm across, in clumps, heart-shaped, irregularly toothed, stalked, white-woolly beneath.
Fruit: achene, pappus forming clock.

BUTTERBUR
Petasites hybridus

The rather stout, large-leaved plants of Butterbur are found in damp places, often growing in masses. Flowering stems with broad bracts appear before the leaves.
Flower: pink or mauve; disc florets only; heads 3-12 mm; sepal-like bracts narrow, purplish.
Flower arrangement: heads in dense racemes. Male and female florets usually on separate plants.
Flowering time: March-May.
Leaf: 10-90 cm across, kidney-shaped, toothed, grey beneath, stalked, up to 1 m high.
Fruit: achene, with pappus.
Uses: the leaves were used to wrap butter, hence the common name.

WALL LETTUCE
Mycelis muralis

This slender, much-branched plant is hairless and grows 20-100 cm high on walls, and in rocky places and woods in calcareous regions.
Flower: pale yellow; 5 ray florets only; heads about 1 cm across; sepal-like bracts in cylinder shape.
Flower arrangement: heads on branched stalks.
Flowering time: July-September.
Leaf: alternate, deeply lobed, coarsely toothed; basal lobes of upper leaves clasping stem; often purple-tinged, red-veined.
Fruit: achene, with pappus.

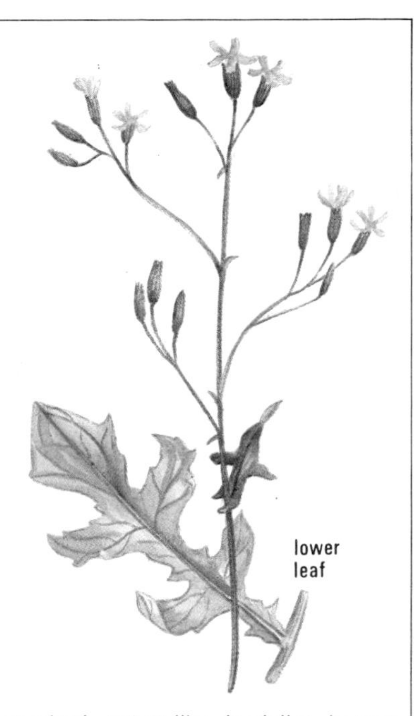

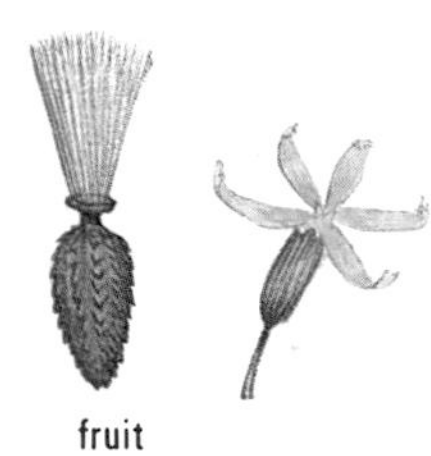

CHICORY
Cichorium intybus

This much-branched plant, with tough, wiry stems, is most commonly seen on roadsides and waste places, mainly on calcareous soils. The stems are grooved and more or less hairy and reach about 120 cm.
Flower: clear sky-blue, 2·5-4 cm across: ray florets only. Carried in small clusters in leaf axils.
Leaf: basal leaves deeply lobed or toothed, rather like dandelion leaves: upper leaves less divided and clasping stem. Often hairy.
Fruit: a ribbed achene: no pappus, but toothed scales on the top.
Uses: Some varieties are cultivated for their stout tap roots, which are roasted and ground and blended with coffee. Other varieties are grown under cover for their succulent basal leaves which, under the right conditions, form compact heads like cos lettuces. Such heads are not found in wild plants.

YARROW
Achillea millefolium

The stem is erect and 8-60 cm high, with soft hairs. Yarrow grows in grass- and waste land and the narrow, ferny leaves may be found creeping through lawns. The plant looks Umbellifer-like, except that the flower stalks do not all join at one point. If crushed, the plant smells aromatic.
Flower: ray florets white or pink, 5; disc cream; head 4-6 mm across.
Flower arrangement: in flat-topped corymbs.
Flowering time: June-August.
Leaf: very finely dissected, fern-like, 5-15 cm long; lower leaves stalked.
Fruit: achene, 2 mm, greyish.

MUGWORT
Artemisia vulgaris

The backs of the finely cut leaves are strikingly white, and the plant is faintly aromatic. The stem is 60-120 cm tall, erect and often red. It grows on roadsides and in waste places.
Flower: brownish-yellow; disc florets only in bell-shaped heads, 3-4 mm long; sepal-like bracts hairy.
Flower arrangement: branched raceme.
Flowering time: July-September.
Leaf: deeply dissected, segments toothed, main vein translucent, almost hairless above, densely white-hairy beneath.

HEMP AGRIMONY
Eupatorium cannabinum

The tall plants grow 30-170 cm high in masses in damp places. The tiny pink flower-heads are gathered in fairly loose clusters.
Flower: pinkish; disc florets only; styles protruding; head of 5-6 florets, 2-5 mm across; sepal-like bracts purple-tipped.
Flower arrangement: heads in rather loose, domed clusters.
Flowering time: July-September.
Leaf: divided to base into 3-5 toothed segments, almost stalkless, opposite; upper leaves undivided.
Fruit: achene, black, with pappus.

TANSY
Tanacetum vulgare

The flower-heads are like little yellow buttons clustered at the ends of stems 30-100 cm high. The plant is highly aromatic and grows in hedgerows and waste places. It is often found as an escape from gardens, where it has been grown for ornament and medicine.
Flower: orange-yellow; disc florets only; heads 7-12 mm across.
Flower arrangement: heads in flat-topped clusters.
Flowering time: July-September.
Leaf: ferny, divided into narrow, toothed segments, alternate, dark green, hairless, gland-dotted, 15-25 cm long.
Fruit: achene, 5-ribbed.

Monocotyledons

FROG-BIT FAMILY Hydrocharitaceae

Members of this family grow partly or wholly submerged in water. Some are rooted, while others float free.

CANADIAN WATERWEED
Elodea canadensis

This waterweed has many little dark green leaves closely spaced up the long stems. It grows submerged in slow-moving fresh water and rarely flowers.
Flower: white, tinged pink, 5 mm across; petals 5.
Flower arrangement: solitary, floating, on long stalk.
Leaf: dark green, translucent, lanceolate, tip pointed or rounded, in overlapping whorls of 3.
Uses: this plant is widely grown in aquaria for the large amounts of oxygen it produces and for the shelter it provides.

WATER-PLANTAIN FAMILY Alismataceae

All the plants in this small family grow near or standing in fresh water. The flowers have 3 white or pink petals rarely lasting more than a day.

COMMON WATER-PLANTAIN
Alisma plantago-aquatica

This is a pretty plant which grows in the wet mud or shallow water of ditches, ponds and slow rivers. The whole plant is hairless and 20-100 cm high. This water-plantain differs from other, more narrow-leaved species in its ovate leaves with abrupt, not tapering bases.
Flower: very pale pink to white, up to 1 cm across; petals 3, rounded; sepals 3; open 1-7 pm.
Flower arrangement: in branched whorls.
Flowering time: June-August.
Leaf: broadly ovate, tip pointed; base joining long stalk abruptly, not tapering.
Fruit: flat seeds in a ring.

ARROWHEAD
Sagittaria sagittifolia

The arrow-shaped leaves and 3-petalled flowers on stout, 3-sided stems poke up out of shallow water. From root to tip the plant is 30-90 cm tall and hairless.
Flower: white, purple blotch at centre, 2 cm across; petals 3, rounded; sepals 3.
Flower arrangement: in short-stalked whorls of 3-5 on long stems. Upper flowers male, lower ones female.
Flowering time: July-August.
Leaf: arrow-shaped, long-stalked, held above water; floating leaves ovate with smaller basal lobes; submerged leaves ribbon-like.
Fruit: in tight, rounded head.

LILY FAMILY Liliaceae

This is one of the largest of the flowering plant families. Many members have bulbs as storage organs. Included in the family are onions, leeks and many garden flowers such as tulips and lilies.

The flowers appear 6-petalled, having 3 true petals and 3 petal-like sepals, which may or may not be joined into a tube. The ovary is superior (above or inside the petal bases).

LILY-OF-THE-VALLEY
Convallaria majalis

This little plant may sometimes be found escaped from gardens, where it is commonly grown. It is also native to Europe in woods, mostly on chalk or limestone. There is a pair of broad leaves at the base of the flowering stem.
Flower: white, sweet-scented, bell-shaped, nodding, 8 mm.
Flower arrangement: in raceme of 6-12 flowers, all directed one way.
Flowering time: May-June.
Leaf: broad, elliptic, stalked, hairless, in a pair; scales sheathing base of leaf-stalks and flowering stem.
Fruit: red berries.
Uses: the plant contains a heart drug close to that found in Foxgloves. It was used to revive gassed soldiers in World War I.

BLUEBELL
Endymion non-scriptus

This species of bluebell is native to the western, Atlantic part of Europe. It was not known to the ancient Greek recorders of plants, hence the species name of *non-scriptus*, meaning 'not written upon'. The plant grows 20–50 cm high, often carpeting woods.
Flower: purplish blue, 1·5-2 cm long, bell-shaped; rim with 6 curled-back teeth; anthers cream; bracts paired, bluish.
Flower arrangement: raceme, drooping at tip; buds erect; open flowers nodding, all directed to one side of long stem.
Flowering time: April-June.
Leaf: long, narrow, all rising directly from bulb, keeled, shiny.
Fruit: a capsule about 15 mm.
Uses: a slimy glue can be obtained by scraping the bulbs.

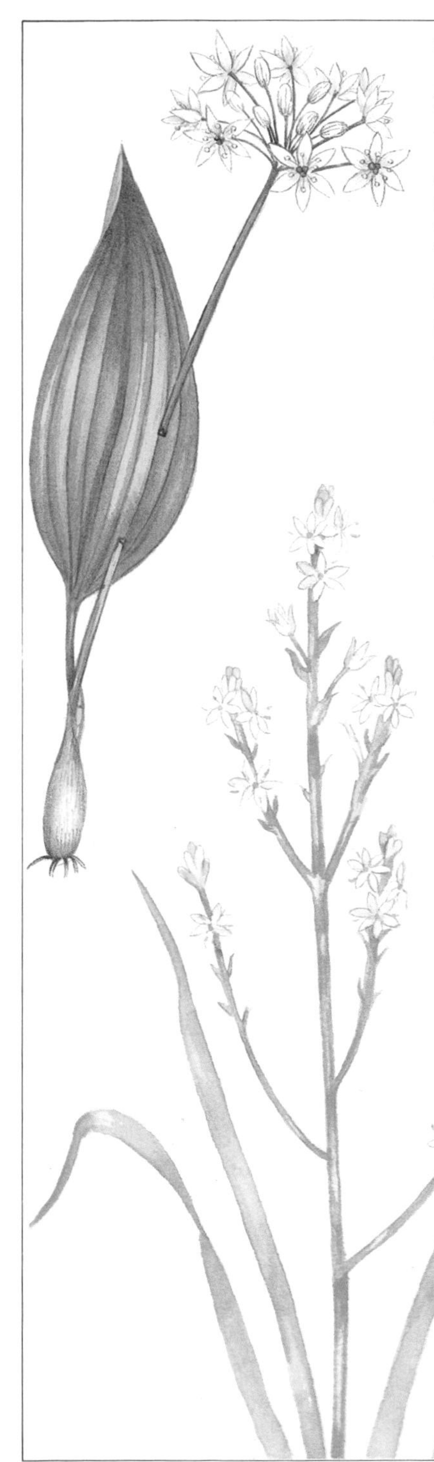

RAMSONS
Allium ursinum

Walkers through woods may notice the strong garlic smell before the plants themselves. Ramsons also grows in hedges. There is a small narrow bulb at the base of the 3-angled stem. The leaves are broad and spring from the bulb.
Flower: white; petals 6, pointed; stamens 6.
Flower arrangement: 6-20 flowers in rounded umbel on unbranched, 3-angled stem 10-45 cm high.
Flowering time: April-June.
Leaf: broad-elliptic, tip pointed, stalked, hairless.
Fruit: 3-lobed, with black seeds.

ASPHODEL
Asphodelus albus

This beautiful and imposing plant is a native of southern Europe, where it flourishes on dry and stony ground, especially on the over-grazed lands around the Mediterranean. It also ascends the southern slopes of the Alps and other mountains to a height of 1600 m, growing in the meadows as well as on the dry scree slopes. Often forming dense clumps, its stout flowering stems reach 1·5 m. There are several similar species.
Flower: white, often with a pink tinge, each petal having a brownish mid-rib. Star-shaped, 3-6 cm across. Borne in dense racemes on leafless branched or unbranched stems. April-July.
Leaf: long and blade-like, V-shaped in cross section: up to 50 cm long and 4 cm broad.
Fruit: a brown capsule with brown seeds.

DAFFODIL FAMILY Amaryllidaceae

This family, centred in tropical and sub-tropical regions, contains many popular cultivated plants. Many are bulbous plants from dry areas. There are 3 petals and 3 petal-like sepals, which may or may not be joined. The ovary is always inferior (below, or outside the petals). The flowers are enclosed in a sheath at first.

SNOWDROP
Galanthus nivalis

Snowdrops are native to Central and S. Europe. In N. Europe they are commonly planted in gardens and widely naturalized in damp woods and hedges. There is a small bulb at the base of the plant, from which spring the leaves and flower stem.
Flower: white and green, bell-shaped, 14-17 mm long; outer petals 3, white; inner petals 3, shorter than outer, deeply notched, green-tipped and striped, forming tube; flower-sheath green.
Flower arrangement: solitary, nodding, on long stem.
Flowering time: January-March.
Leaf: linear, grey-green, keeled.
Fruit: capsule.

WILD DAFFODIL
Narcissus pseudonarcissus

The Wild Daffodil is native in woods and grassland. The flower stem and leaves end in a bulb of 2-3 cm.
Flower: trumpet golden yellow, as long as or slightly shorter than surrounding 6 paler yellow petals, 3·5-6 cm long; flower-sheath brown, papery.
Flower arrangement: solitary, drooping, on flattened stalk 20-35 cm high.
Flowering time: February-April.
Leaf: linear, 12-35 cm long, grey-green.
Fruit: capsule.

IRIS FAMILY Iridaceae

Many members of this family, including crocuses, gladioli, and the various kinds of irises, are grown in gardens. A large proportion of the thousand or so species live in hot and dry areas. Many have underground storage organs in the form of corms or rhizomes. The 3 petals and 3 petal-like sepals, which may or may not be of similar shape, are joined into a tube at the base.

YELLOW IRIS
Iris pseudacorus

The leaves and flower-stem are tall and stiff, growing 40-150 cm high in river or ditch margins, marshes or marshy woods. The thick rhizomes often arch up above the soil surface.
Flower: yellow, 8-10 cm across; outer petals 3, broad, drooping, often dark-veined; inner petals 3, erect; stigmas 3, petal-like, forked at tip.
Flower arrangement: 2-3 on long, flattened stem.
Flowering time: May-July.
Leaf: very long, narrow, blue-green, 1·5-2·5 cm across, midrib raised; together, leaves form a flattened fan-shape at base.
Fruit: capsule; seeds brown.

PURPLE CROCUS
Crocus vernus

One of several rather similar species, this plant is a native of the mountains of southern and central Europe, often pushing up its delicate flowers before the snows have melted from the meadows. Introduced and naturalized in grasslands elsewhere in Europe.
Flower: purple or violet, to 6 cm high. 6 petals, opening widely in sunshine. 3 stamens and a feathery orange stigma. The 'stalk' is the base of the petal tube. February-June.
Leaf: bristle-like, 4-5 cm long.
Fruit: a capsule at ground level.

YAM FAMILY Dioscoreaceae

Most members of this family are tropical climbers with tuberous rhizomes. Most are herbaceous, but a few are woody. Several species, known as yams, are cultivated in the tropics for their starch-filled tubers. About 6 species occur in Europe. Unlike most monocotyledons, the leaves are net-veined.

BLACK BRYONY
Tamus communis

The only widespread European species in the family, this clockwise-twining herbaceous climber is common in hedgerows and woodland margins. Its slender, hairless stems reach heights of 4 m. New stems spring from the rhizome each year. Absent from the north.

Flower: greenish-yellow, funnel or star-shaped, 4-5 mm across. 3 sepals and 3 petals alike and joined into a tube at the base. Borne in slender, loose racemes in leaf axils. Male and female flowers on separate plants. May-August.

Leaf: heart-shaped, to 10 cm long on long curving stalks: deep glossy green.

Fruit: clusters of shiny red berries. Poisonous.

(See also White Bryony – page 76.)

ORCHID FAMILY Orchidaceae

This very large family, with over 20,000 species, is famed for the beauty and intricacy of its flowers and many tropical species are cultivated commercially. There are 3 sepals, usually more or less alike and often brightly coloured, and 3 petals. The lowest petal, known as the lip or labellum, is very different from the others – often of bizarre shape and pattern, and remarkably like an insect in the bee orchid and its relatives. It is often prolonged as a nectar-filled spur. The elaborate shapes and colours, combined with scent and nectar, are designed to attract pollinating insects, but they are often so specialized that only one kind of insect can pollinate them. The pollen is normally in sticky clumps. The seeds are minute. Several species, such as the Bird's Nest Orchid, are saprophytes – devoid of chlorophyll and obtaining food from dead leaves. Most orchids flower sporadically and should never be picked.

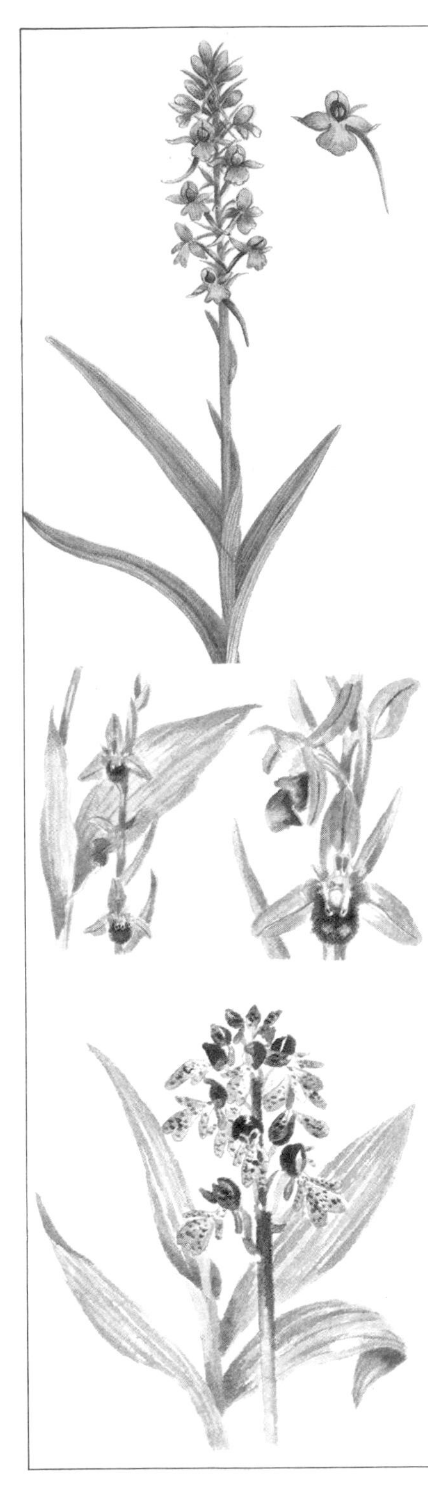

FRAGRANT ORCHID
Gymnadenia conopsea

The scent of this orchid is strongest in the evening, when it attracts pollinating moths. The spurs on the flowers are very narrow and long. 15-40 cm high in grassland and marshy ground, particularly on chalk or limestone.
Flower: pink-lilac; lip of 3 rounded lobes; spur 11-13 mm long, slender; bracts equalling flowers.
Flower arrangement: dense spike.
Flowering time: June-August.
Leaf: long, narrow, folded down midrib, tip more or less rounded, slightly fleshy; bases sheathing stem.
Fruit: capsule.

BEE ORCHID
Ophrys apifera

The lip of this fascinating orchid looks like a bee. It even smells like a female bee and in some parts of Europe it is pollinated by male bees which try to mate with it. Elsewhere it is normally self-pollinated. Reaching 40 cm, it is not uncommon in grassland and on rough ground on calcareous soils.
Flower: sepals pink (occasionally white), oval to 1·5 cm long: 2 upper petals greenish or pink, usually narrow and blunt-ended: lip rounded, rich brown and furry with pale U or W mark, to 1·5 cm long. Up to 10 well-spaced flowers on a spike. May-July.
Leaf: pale green, ovate to lanceolate: basal leaves form rosette which withers as flowers appear.
Fruit: a ribbed capsule.

LADY ORCHID
Orchis purpurea

Each flower of this beautiful plant resembles a lady in a poke bonnet and broad skirt. Although very rare in Britain, it is not uncommon elsewhere in open woods and scrub on calcareous soils. It reaches 50 cm.
Flower: 3 sepals form a purplish-brown hood or bonnet enclosing 2 upper petals. Lip, about 1·5 cm long, is white or pale pink with tufts of tiny crimson hairs; 3-lobed, the middle one broad and skirt-like, sometimes split up the middle. In dense spike. April-July.
Leaf: bright green, to 15 cm long: mostly in a basal rosette.
Fruit: a capsule with many seeds.

COMMON SPOTTED ORCHID
Dactylorhiza fuchsii

This conspicuous orchid is also one of the more common, and grows 15-50 cm high in damp or dry grassy places and open woods. Like many other orchids, this one hybridizes readily with other species.
Flower: pale pink or whitish with dark red marks; segments each side of hood pointing up; lip broad, 3-lobed; middle lobe pointed, slightly longer than broader side lobes; spur 5·5-8·5 mm long; bracts equalling or longer than flowers.
Flower arrangement: dense spike.
Flowering time: June-August.
Leaf: usually dark-blotched; lower leaves broad-elliptic, tip rounded.
Fruit: capsule.

EARLY PURPLE ORCHID
Orchis mascula

These spring-flowering orchids grow 15-60 cm high in moist woods and grassland. They often grow among bluebells and Dog's Mercury. The lovely purple or mauve flowers have the unlovely smell of tom cats.
Flower: purple or mauve; lip pale with dark spots at top-centre, broad, 3-lobed; spur thick, blunt; bracts purple.
Flower arrangement: spike.
Flowering time: April-June.
Leaf: usually dark-spotted, alternate, slightly fleshy, oblong, tip shortly pointed or rounded; bases sheathing stem.
Fruit: capsule.
Uses: because of the twin tubers, looking like testicles, this orchid was used as an aphrodisiac. The name orchid comes from the Greek for testicle. Most European orchids grow from tubers.

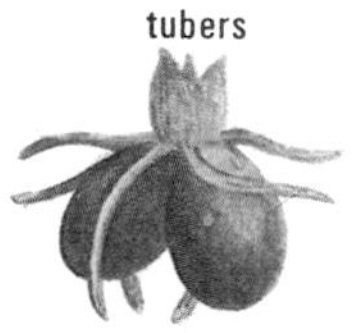

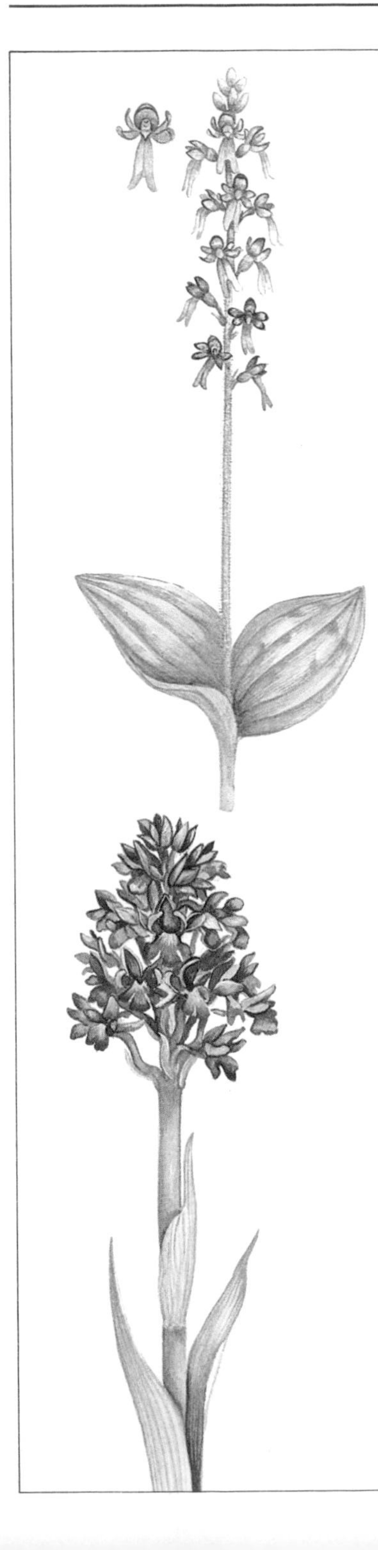

COMMON TWAYBLADE
Listera ovata

The common name refers to the 2 broad leaf-blades just below the middle of the stem. The stem is 20-60 cm high and hairy above the leaves. The plant grows in damp woods and grassland, particularly on limestone or chalk, and seems equally happy in shade or bright sunlight. The plant has slender roots without tubers. Lesser Twayblade (*L. cordata*) is a much smaller plant with reddish flowers. It grows in pine woods and on peat bogs and moorland, where it flowers June-September.

Flower: yellow-green; lip oblong, furrowed down centre, forked at tip; no spur; bracts tiny.

Flower arrangement: long spike. with few to many flowers: up to 25 cm long.

Flowering time: May-July.

Leaf: pair of very broad, ovate leaves, slightly fleshy; bases sheathing stem.

Fruit: capsule.

PYRAMIDAL ORCHID
Anacamptis pyramidalis

Named for the markedly pyramidal shape of the young flower spike, this beautiful plant grows on dry, grassy slopes, especially on calcareous soils, and on shelly sand dunes by the sea. Rising from 2 oval tubers, it reaches a height of about 50 cm.

Flower: pale to deep pink and usually unspotted; lip about 5 mm long and deeply 3-lobed with a long curved spur. 2 parallel ridges at base of lip guide insect tongue past the pollen masses as it probes the spur for nectar. Flowers borne in dense pyramidal clusters at first, becoming oblong with age. May–August.

Leaf: narrow; greyish green and unspotted: 5-8 in a rosette.

Fruit: a capsule with many seeds.

BROAD-LEAVED HELLEBORINE
Epipactis helleborine

This helleborine has up to 3 stems, often tall, short-hairy at the top. They rise from a short length of underground stem and grow up to 80 cm high in hedgerows and woods. The cup-shaped lip of the flower holds nectar for bees and wasps.
Flower: green to dull purplish, scentless; lip cup-shaped with narrow, turned-back tip; no spur; bracts almost equalling flowers.
Flower arrangement: loose spike of drooping flowers turned to one side.
Flowering time: July-October.
Leaf: alternate, broadly ovate, tip pointed, middle leaves longest, margins rough.
Fruit: capsule.

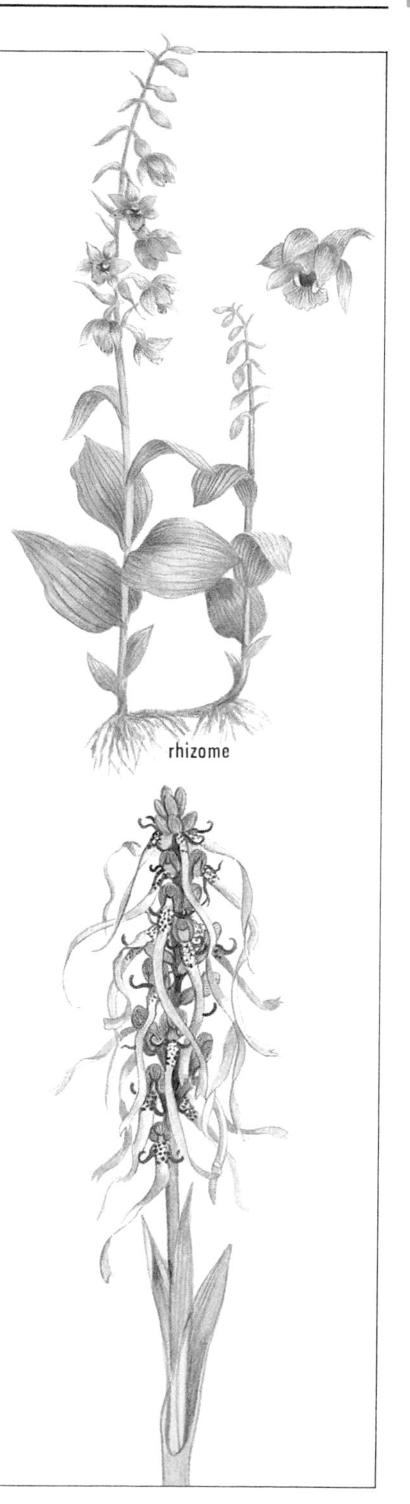

LIZARD ORCHID
Himantoglossum hircinum

Named for the long, ribbon-like lip, this is one of Europe's largest orchids, occasionally reaching 1 m. It grows mainly on dry, calcareous grassland: very rare in Britain, but locally common in southern and central Europe.
Flower: sepals and 2 upper petals greenish, forming a hood; lip cream, fawn, or pinkish with darker spots, divided into 3 ribbon-like lobes of which the central one reaches 5 cm. Smells of goats. Carried in long spikes. May-July.
Leaf: pale greyish-green, lanceolate, up to 20 cm: 6-8 root leaves in a rosette, withering as flower stem grows. Upper leaves sheathe stem.
Fruit: a many-seeded capsule.

ARUM FAMILY Araceae

Members of this large, mainly tropical family include Arum Lilies of florists' shops (not lilies at all) and the Swiss Cheese Plant (*Monstera deliciosa*), a common house-plant. The poker-like flower-spike is surrounded by a large sheath; many exude unpleasant smells to attract pollinating flies.

LORDS-AND-LADIES
Arum maculatum

The tiny flowers are hidden at the base of a purple, poker-shaped stalk, half-enclosed by a yellow-green hood. Midges are trapped in the base of the hood and pollinate the flowers. They later escape and effect cross-pollination by becoming trapped in other flower-hoods. The plants grow in shady hedges and woods. 15-50 cm high.
Flower: unisexual, without petals or sepals.
Flower arrangement: clustered at base of spike: female flowers at bottom, then male flowers, and then a ring of sterile flowers forming a ring of hairs. Insects can push down past the hairs, but cannot get out again. Female flowers open first; male flowers then open and dust insects with pollen; hairs then wither and insects escape to carry pollen to other spikes.
Flowering time: April-May.
Leaf: arrow-shaped, usually dark-blotched, shiny; stalks rising directly from roots.
Fruit: spike of orange berries.

DUCKWEED FAMILY Lemnaceae

The plants in this family are very small and made up of simple, leaf-like lobes which float on fresh water. Included in the family are species of *Wolffia* – the smallest known flowering plants.

DUCKWEED
Lemna minor

The tiny, round, leaf-like lobes of this plant spread on the surface of still water, often covering large areas. A single thread-like root hangs in the water from each plant. The plants normally reproduce by the budding of new lobes, and over-winter in the mud of pond- or ditch-bottoms. Each lobe is 1·5-4 mm across.
Flower: minute; unisexual with just 1-2 stamens or 1 carpel. Carried in a pouch on lobe margin, but rare.
Flowering time: June-July.

BUR-REED FAMILY Sparganiaceae

Some plants in this small family float in fresh water, while others grow erect at the water-side. The tiny flowers are gathered into round heads on the same stem, male separate from female. Most species are found in temperate regions.

BRANCHED BUR-REED
Sparganium erectum

This is a stout, erect, hairless plant of 50-150 cm, growing at the edge of still or slow-moving fresh water. The long, stiff leaves look like iris leaves except for the 3-angled bases. The less common Unbranched Bur-reed (*S. emersum*) has a single stem of flower-heads.
Flower: tiny, with 6 greenish, black-edged papery scales.
Flower arrangement: males and females in separate, round heads on branched, leafy stem (male above female).
Flowering time: June-August.
Leaf: long, narrow, 3-angled and sheathing stem at base.
Fruit: in bur-like head, each bearing spike.
Uses: stands of Bur-reed shelter wild-fowl, and the fruits provide food for them in winter.

REEDMACE FAMILY Typhaceae

Most species in this small family are very tall and grow in shallow, fresh water. The tiny flowers are packed into long, dense spikes or clubs. The plants are commonly known as 'Bulrushes'.

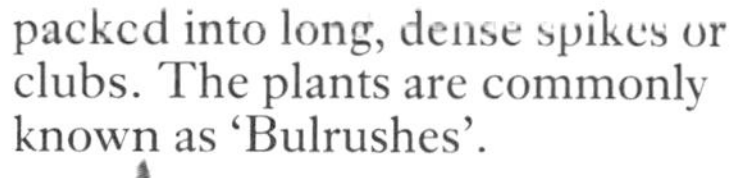

GREAT REEDMACE
Typha latifolia

Great Reedmace is very tall (1·5-2·5 m), and grows in large stands in the shallows of still or slow-moving fresh water. It is often found growing with Reed (*Phragmites communis*), a tall water-side grass.
Flower: tiny and unisexual. No petals but surrounded by hairs – brown in female, yellow in male.
Flower arrangement: females in large, dense club-shape, continuous with thinner, paler male spike above (Lesser Reedmace, *T. angustifolia,* has a gap between).
Flowering time: June-July.
Leaf: very long, narrow, 10-18 mm wide, grey-green, rather leathery.
Fruit: wind-borne achene.
Uses: the leaves can be used for weaving and basketry.

RUSH FAMILY Juncaceae

The rushes are often confused with grasses, but they are not closely related. Rushes are slender herbs, usually with creeping rhizomes, and they bear small and inconspicuous flowers, although these may be gathered into large bunches. There are 6 perianth segments (3 sepals and 3 petals), all alike and usually greenish white or brown, with a chaffy texture. There are usually six stamens and three stigmas, and the flowers are wind-pollinated. True rushes (*Juncus* species) usually have hairless and rather spiky cylindrical leaves, while woodrushes (*Luzula* species) have flat and hairy leaves. Most true rushes flourish in damp places and many of the 300 or so species grow in cold climates.

JOINTED RUSH
Juncus articulatus

This rather weak-stemmed rush is named because the leaves contain transverse partitions, which can be felt by pulling a leaf between the fingers. (Several other rushes also have these partitions.) The stem is dark green and smooth and reaches 80 cm. It grows on wet ground almost everywhere.
Flower: dark brown with pointed segments. In loose, much-branched heads at top of stems. June-October.
Leaf: deep green, usually curved and slightly compressed: 2-7 on a stem.
Fruit: a many-seeded capsule.

FIELD WOODRUSH
Luzula campestris

Abundant in grassy places almost everywhere, this grass-like little plant grows in loose tufts and reaches about 15 cm.
Flower: chestnut brown with conspicuous yellow stamens: star-like, 3-4 mm across. Carried in a loose panicle. March-June.
Leaf: bright green, to 4 mm wide: thinly clothed with long hairs.
Fruit: a 3-seeded capsule.

SOFT RUSH
Juncus effusus

This very common rush grows in tufts in wet places almost everywhere. The glossy green stems are quite smooth, up to 3 mm thick and reaching 1·5 m in height.
Flower: brown with narrow segments. Sprouting in a cluster in upper part of stem. The cluster may be tightly packed or loosely branched. June-August.
Leaf: reduced to reddish brown scales at base of stem.
Fruit: a many-seeded capsule.

SEDGE FAMILY Cyperaceae

Members of this family are grass-like herbs, but they differ from grasses in having solid, unjointed stems, often triangular in cross-section. The slender leaves are also arranged in a triangular fashion around the shoots. The flowers are very small, without petals, and grouped into brown spikes. In the true sedges (*Carex* species) the flowers are unisexual and the male and female flowers often form separate spikes. They are wind pollinated. The fruits are like little grains – globular or distinctly 3-cornered – and often important for identifying the species. Most of the 4,000 or so sedges flourish in wet places, but several small species, such as the bluish-leaved Glaucous Sedge (*Carex flacca*), also grow in dry grassland.

COMMON COTTONGRASS
Eriophorum angustifolium

Despite its name, this wiry plant is a sedge. Reaching 60 cm, it is abundant on wet heaths and bogs, forming extensive patches with the aid of its creeping rhizomes and forming white drifts as the fruits ripen.
Flower: very small, hermaphrodite: no petals, but each flower has a ring of hairs which become long and cottony after flowering – hence the common name. Flowers grouped into several nodding spikes. June-August.
Leaf: deep green, up to 20 mm long and 6 mm wide; deeply channelled. Becoming brown and dying by flowering time, to be replaced by fresh green leaves in late summer.
Fruit: a 3-cornered nutlet, surrounded by the long white hairs.
Harestail Cottongrass (*E. vaginatum*) has its flowers grouped into a single oval spike. It is common on moorland.

GREAT POND SEDGE
Carex riparia

This is a typical waterside sedge, growing in dense clumps around ponds and slow-moving streams. The robust stems are stiff and hairless and distinctly 3-cornered. They reach 1·5 m.
Flower: unisexual, with male and female flowers in separate spikes. Up to 7 male spikes closely grouped at top of stem; dark brown, becoming yellow as stamens open and shed pollen. Up to 5 female spikes lower down, greenish and up to 9 cm long; lower ones often nodding. Long tapering bracts below each spike. May-June.
Leaf: bluish-green, up to 1·5 cm wide, sharply keepled and with sharp edges: often taller than stems.
Fruit: an ovoid or 3-cornered nutlet.

GRASS FAMILY Gramineae

With over 10,000 species, this is one of the most important plant families – providing the staple diet of many wild and domestic animals and also giving us all our cereals. Grasses grow in all parts of the world and in all kinds of habitats. Most are small herbaceous plants with hollow, jointed stems, although the woody bamboos are also grasses.

The flowers have no petals and consist of a single carpel and 3 stamens. They are wind-pollinated. They are enclosed in spikelets, which consist of two outer scales called glumes and a number of smaller inner scales. There may be one or more flowers in a spikelet, and the spikelets may be borne in dense spikes (ears) or in spreading panicles. Most species flower May–July. The lower parts of the leaves ensheathe the stem, and where they join it there is a little membranous collar called a ligule. The shape of this ligule helps to identify grasses. The fruit is normally a 1-seeded grain.

WALL BARLEY
Hordeum murinum

This common annual of waste places and disturbed ground reaches 60 cm in loose tufts.
Leaf: light green, to 20 cm long: finely pointed. Ligule very short and ring-like.
Spikelet: with fine, stiff bristles (awns): borne in '3's on a somewhat flattened bristly spike which children like to throw like darts. Only the middle one of each trio has a flower.

QUAKING GRASS
Briza media

This delicate grass occurs in all kinds of grassland, including roadside verges. It grows in loose tufts and reaches 75 cm. It gets its name for its drooping and freely-swaying spikelets.
Leaf: bright green, hairless, to 15 cm. Ligule up to 1·5 mm: blunt-ended.
Spikelet: flattened, oval or triangular to 7 mm long. Shining brown, often purplish, with up to 12 flowers. In a loosely branched, pyramidal panicle.

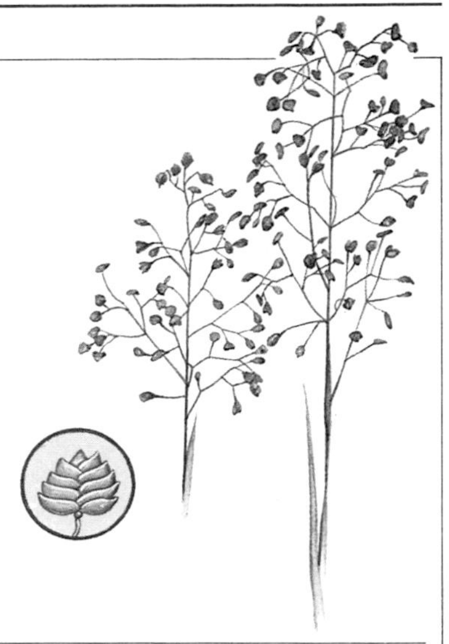

PERENNIAL RYE GRASS
Lolium perenne

This very nutritious, but rather tough grass is extensively sown for pasture and for lawns, where its wiry flower stems often defeat the mower. It reaches about 1 m when not cut or grazed. There are many varieties.
Leaf: bright green and hairless, to 20 cm long; folded along mid-rib when young. Basal sheaths often pink. Ligule ring-like, to 2 mm high.
Spikelet: stalkless, oval and flat, with only 1 glume and up to 14 flowers. Placed alternately on opposite sides of a slender spike, with narrow edge facing the stem.

COCKSFOOT GRASS
Dactylis glomerata

Named for the resemblance of the panicle to a chicken's foot, this grass forms dense clumps in all kinds of rough grassland. It is nutritious and regularly sown for hay. It reaches 1·5 m.
Leaf: pale or greyish green to 45 cm: folded down centre when young. Ligule to 12 mm long: more or less triangular.
Spikelet: to 9 mm long with 2-5 flowers: in dense clusters, all facing one way on a few-branched panicle.

COMMON REED
Phragmites communis

Forming dense stands in fens and marshes and around the edges of lakes and rivers, this grass reaches heights of 3 m. Its tough stems are in great demand for thatching.

Leaf: greyish green, to 60 cm or more long and 3 cm wide, with a fine tapering point. Ligule is a fringe of tiny hairs.

Spikelet: up to 15 mm long, with 2-6 flowers; very hairy. Grouped into dense purplish or brownish feathery panicles.

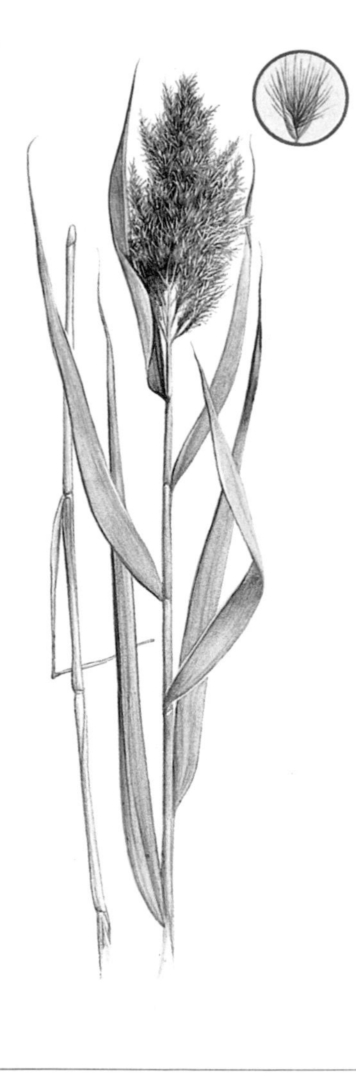

MEADOW FOXTAIL
Alopecurus pratensis

This grass is especially common on rich, moist soils, particularly in riverside meadows. It reaches 1·2 m and is highly nutritious. It is one of the earliest-flowering grasses.

Leaf: mid-green, tending to bluish; hairless but often slightly rough. Ligule more or less square, to 2·5 mm long.

Spikelet: flattened, oval, 4-6 mm long: 1-flowered and hairy, with a straight bristle. Packed into dense cylindrical spikes to 13 cm long, greyish-green or purplish and very soft to the touch.

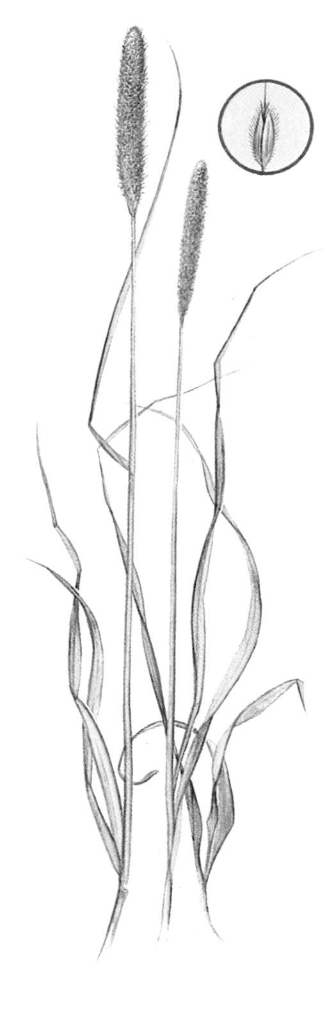

WAVY HAIR GRASS
Deschampsia flexuosa

This slender grass forms neat cushion-like tufts on heaths and moorlands. Its reddish stems reach 1 m.
Leaf: bright green and hairless; tightly in-rolled and bristle-like to 20 cm or more long. Ligule conical, to 3 mm long and bluntly pointed.
Spikelet: up to 6 mm long, usually 2-flowered. Shiny brown or purple. Borne on wavy, hair-like stalks in loose panicles.

YORKSHIRE FOG
Holcus lanatus

This is a very furry grass of roadsides, waste ground, and run-down pastures. It reaches 1 m.
Leaf: greyish-green and clothed with soft hair; to 20 cm long and finely-pointed. Ligule flat-topped, to 4 mm.
Spikelet: to 6 mm long, 2-flowered; white, green, pink, or purple. Borne in rather dense hairy panicles, lanceolate at first but opening to a pyramidal shape.

TIMOTHY GRASS
Phleum pratense

Reaching 1·5 m, this loosely tufted and very nutritious grass is characteristic of low-lying meadows, but is sown extensively for hay and pasture. It also grows on many roadsides.
Leaf: green or grey-green, up to 45 cm long; hairless, but rough to touch, at least on upper surface. Ligule up to 6 mm long, oval with blunt tip.
Spikelet: oblong, flat, 3-4 mm long; 1-flowered, with 2 short bristles. Packed into a dense cylindrical spike 6-30 cm long; greyish-green or purple and rough to the touch.

COMMON BENT GRASS
Agrostis tenuis

One of several very similar grasses, this species is abundant almost everywhere, but especially on the drier acidic soils of heathlands and on the rough grazings of the uplands. It reaches about 70 cm and its delicate flower heads form a brown mist over the ground.
Leaf: bright green and hairless, to 15 cm long, finely pointed and often in-rolled. Ligule ring-like to 2 mm.
Spikelet: 2-4 mm long, 1-flowered, green at first becoming purplish-brown. Borne on spreading branches on a rather pyramidal panicle.

FEATHER GRASS
Stipa pennata

This beautiful grass grows on dry, stony slopes in the Alps, reaching altitudes of 2,500 m. Up to 1 m in height, it is quite unmistakable because of its long plumed awns.
Leaf: blue-green, to 50 cm or more; bristle-like and in-rolled, especially in dry weather.
Spikelet: to 2 cm long, pale brown when ripe; 1-flowered with a feather-like awn to 35 cm long. A small cluster of spikelets at top of stem. The awn carries the fruit away, and twisting movements in response to humidity changes drive the pointed grain into the ground.

ROUGH-STALKED MEADOW GRASS
Poa trivialis

Abundant in meadows and pastures, especially on the moister soils, this grass also occurs on waste land and as a weed of cultivation. Loosely tufted and up to 1 m high, it is one of several similar species, but distinguished by its rough stems.
Leaf: deep green, often with a purplish tinge, to 20 cm long and abruptly pointed. Ligule 4-10 mm long, parallel-sided with a triangular tip.
Spikelet: ovate to 4 mm long; 2-4 flowers; dull green, often with a purplish tinge. Borne on a distinctly pyramidal panicle.

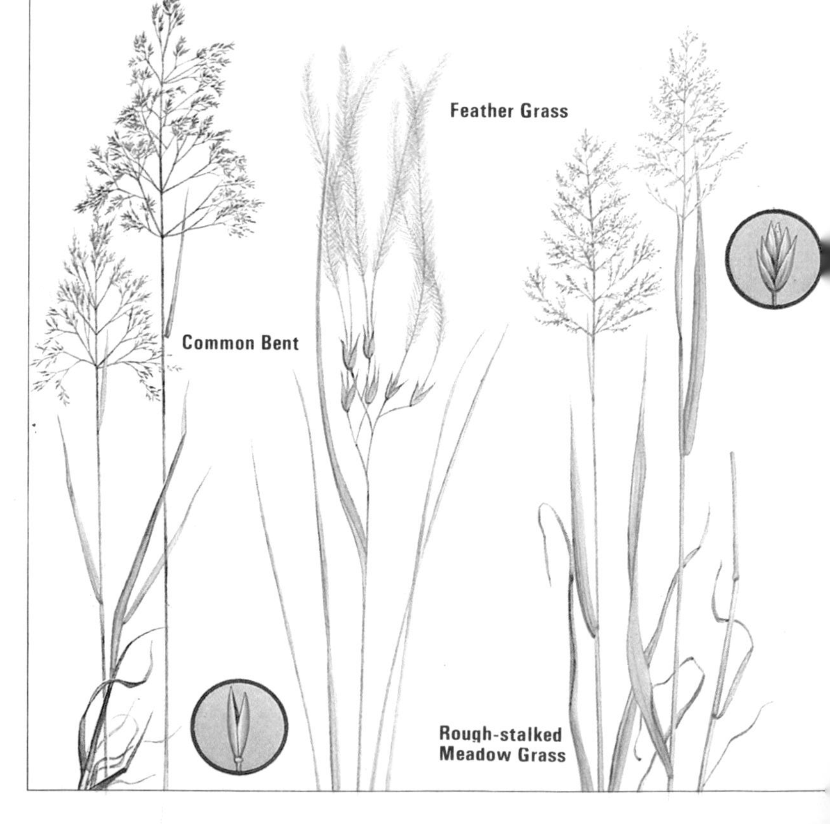

MARRAM GRASS
Ammophila arenaria

This grass grows on coastal sand dunes, where its extensive creeping rhizomes help to bind and stabilize the sand.
Leaf: greyish-green to 60 cm long; tightly in-rolled. Ligule to 3 cm long, very narrow and pointed.
Spikelet: to 16 mm long, 1-flowered. Packed into dense, pale spike-like panicles, to 22 cm long.

FALSE OAT GRASS
Arrhenatherum elatius

Abundant on roadsides and other rough grassland, this loosely-tufted grass is very attractive as its shiny heads sway in the breeze. It reaches 1·5 m.
Leaf: bright green, rough textured and usually lightly haired; to 40 cm long. Ligule to 3 mm long, smoothly rounded at end.
Spikelet: up to 11 mm long, usually 2-flowered and with 1 or 2 long bristles (awns). Shiny, green at first becoming brown or purplish. Panicle slender at first, but branches – in well-spaced clusters – later spreading and nodding.

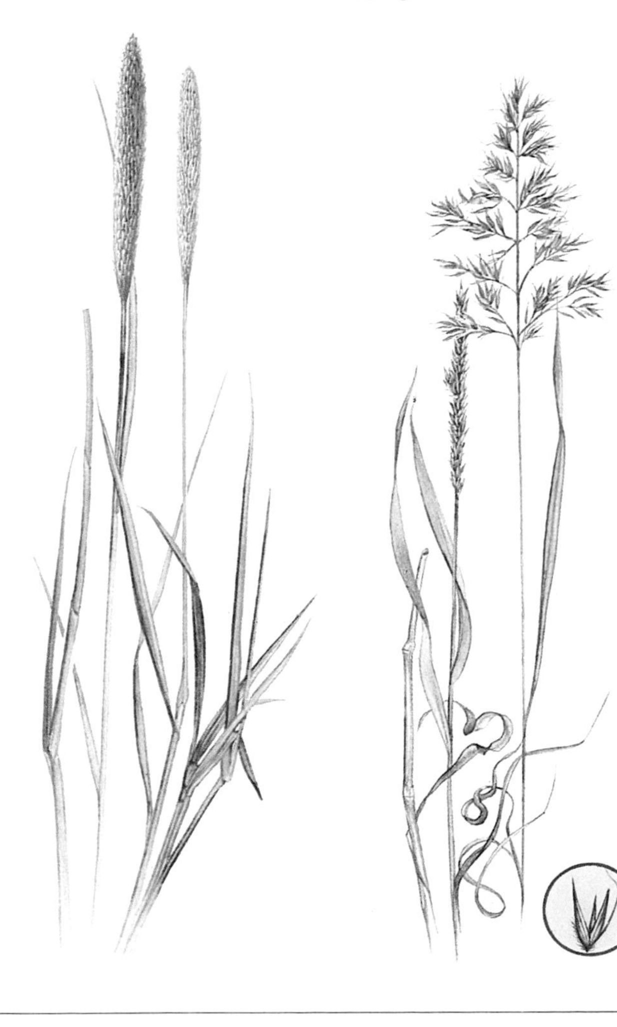

CHAPTER TWO

TREES AND SHRUBS

Trees and shrubs are plants with long-lived woody stems. A tree normally has one main stem, known as the trunk, while a shrub usually has several main stems all arising at or near ground level. Shrubs are rarely more than about 5 metres high. A tree which is cut to ground level may send up several new trunks and thus appear shrub-like for a few years, but it does not become a true shrub. Any fairly low dense woody plant may also be called a bush, but this is not a precise term and it is applied equally to true shrubs and to trees which have been cut or pruned into a shrub-like shape. The shrubs included in the following pages all reach at least a metre in height, although there are many low-growing species, including heathers and their relatives. Some of the latter 'creep' along within a couple of centimetres or so of the ground. The low-growing species, often known as under-shrubs, are included in the previous chapter.

Flowers and Cones

Most of the trees and shrubs are flowering plants (Angiosperms), and, apart from the fact that they usually live longer, their life histories are identical to those of the smaller, herbaceous plants. Even their flower structures are the same. There are, however, a number of cone-bearing species among the trees and shrubs. These belong to the group known as Gymnosperms and they are slightly less advanced than the Angiosperms.

Gymnosperm pollen is carried in small male cones and blown to the female cones by the wind. Seeds later develop in the woody female cones, but they are never completely enclosed in fruits as they are in the flowering plants. The seeds may take up to two years to reach maturity. Male and female cones may be borne on the same tree or on different trees. The flowering trees, the Angiosperms, often carry male and female organs in one flower, but there are many species, including the oaks and the birches, which have single-sexed flowers. Like the cones, these male and female flowers may grow on the same tree or on different trees.

Evergreen and Deciduous

Species that retain their leaves throughout the winter are called evergreens, while those that drop their leaves in the autumn, often after beautiful colour-changes, are known as deciduous trees. The autumn colours of the deciduous trees are caused by chemical changes, triggered off by falling temperatures and related to the withdrawal of useful food materials from the leaves before they fall.

Most conifers are evergreen, while most European flowering trees are deciduous. The flowering trees are also known as broad-leaved trees, because their leaves are generally much broader than the thin needle-like leaves of the conifers. Evergreen trees and shrubs grow mainly in the colder regions, though many evergreen species, such as the Cork Oak and the Stone Pine, are found in the Mediterranean region where the summers are very hot and dry. The tough leaves of the evergreens have waxy coats and other features which cut down water loss and enable them to withstand not only winter frosts but also summer droughts – both of which make absorbing moisture from the ground difficult.

Identification

Tree shape can often provide the first clues to identification. When a tree grows alone and has space to develop properly, its general shape, and especially its crown of branches, is often characteristic of its species. But shape cannot be relied on to identify species that grow in woodland where neighbouring trees interfere with growth and alter form. Many trees have very distinctive bark and this can be used to identify them at all times of the year; look for the ways in which the outer layers of the bark crack or flake. The shape and colour of winter buds can also be used to identify many deciduous trees, while the colour of the shoots – the youngest parts of the branches – is a useful guide when trying to identify certain shrubs.

Flowers, although they are present only at certain times of the year, afford the surest way of identifying most of the flowering trees. Many trees carry their flowers in catkins – elongated clusters of unisexual flowers which may hang down or stand erect on the branches. Cones, too are of great help in identification, as they vary a good deal in shape according to the species. Again, some hang from branches, while others stand upright; many break up when mature, while others stay whole and may remain on the tree for several years after scattering their seeds.

Leaf shape is another very useful guide, and some of the main shapes of deciduous leaves are shown here. The margin may be quite smooth (entire), but is normally toothed to some extent and often deeply lobed. Compound leaves are divided into several distinct leaflets. With the conifers, it is the way in which the needles are arranged on the shoots that is particularly important: some are attached singly and some in small whorls, while all the pines carry their needles in small bunches, with two, three or five needles in each bunch. Most European pines have just two needles in each bunch.

Fruits can also aid identification, although several different kinds of fruits can be found in one family. The nut, typified by the acorn and the hazel nut, is a dry fruit with a woody coat and a

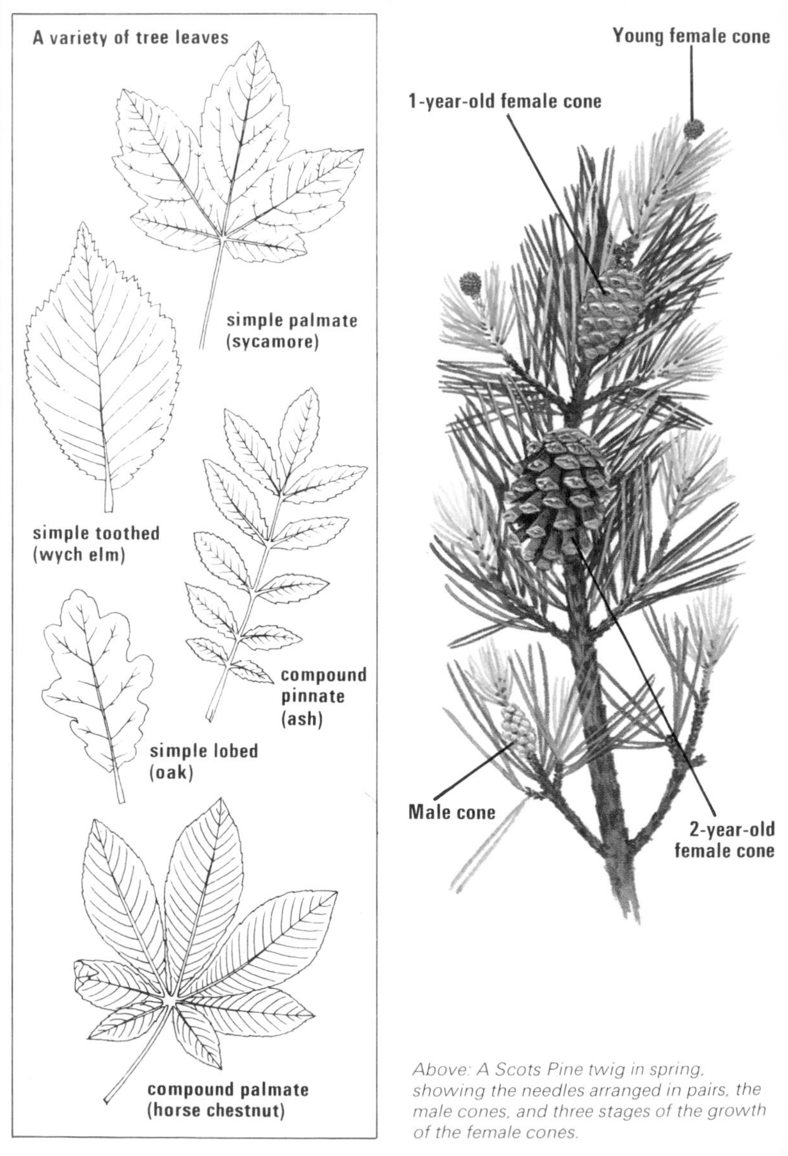

Above: A Scots Pine twig in spring, showing the needles arranged in pairs, the male cones, and three stages of the growth of the female cones.

single seed. Berries are juicy fruits with several seeds; the skin is usually thin, though it is quite thick in oranges and their relatives. Drupes or stone fruits have a woody inner layer (the stone) surrounding a single seed. The outer part of the fruit is usually fleshy, as in cherries and plums. The walnut, despite its name, is also a drupe, but its outer part is leathery. Apples and pears are known as pomes; the edible part is the swollen top of the flower stalk in which the seeds are embedded.

Measuring Trees

The height of a tree rarely has much bearing on its identification, but it is often of interest to know just how high certain trees are. Here is a simple method of estimating height. You will need a straight stick the same length as the distance from your eye to your outstreched hand. Hold the stick vertically before you at arms length. Walk backwards or forwards in front of the tree until the top and bottom of the stick coincide with the top and bottom of the tree. The distance from where you are standing to the base of the tree is then equal to the height of the tree.

MEASURING TREE HEIGHT

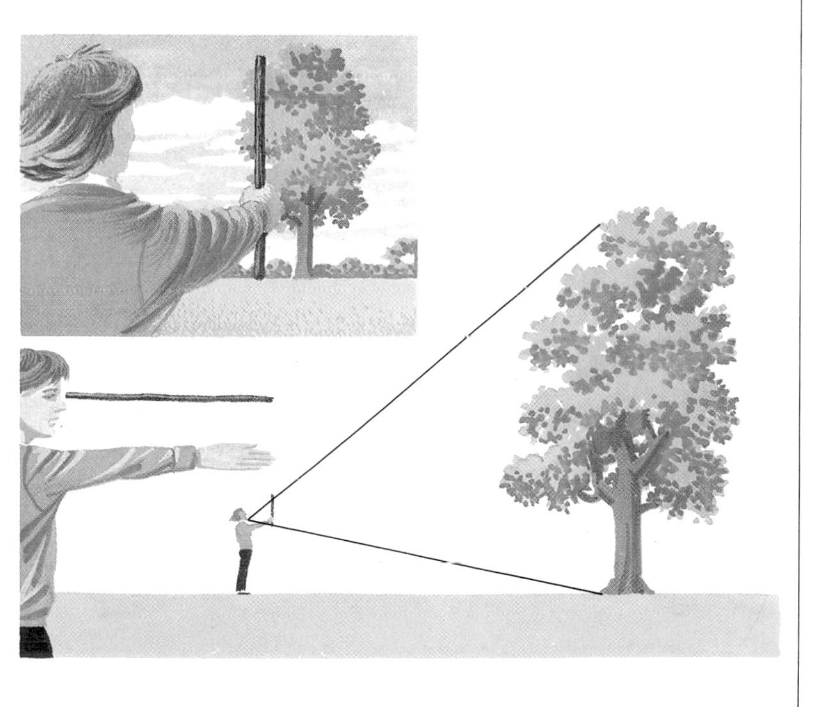

Cone-bearing Trees

PINE FAMILY Pinaceae

This important family of over 300 species occurs in the Northern hemisphere, forming forests in the cooler northern regions and restricted to mountains in the south. The trees belonging to it – firs, cedars, larches, spruces, pines, and hemlocks – all have needle-like leaves and woody cones made up of spirally arranged scales. Nearly all are evergreens.

SILVER FIR
Abies alba

This European conifer forms pure forests in mountainous regions of central Europe, from the Pyrenees across the Alps to the Balkan mountains. It is widely cultivated for its timber and other useful products and it grows to 57m.

Crown: narrow and conical, with level branches that have upturned tips.

Bark: smooth, dark grey, and blistered in young trees, becoming cracked into small square plates with age.

Shoots: grey-buff, with dark hairs.

Buds: red-brown and egg-shaped.

Leaves: thick needles with notched tips, dark green above with two narrow white bands beneath; 2–2·5cm long. They are arranged in two rows along the twigs.

Cones: erect and cylindrical, 10–15cm long, ripening from green to orange-brown. The large scales have bracts turned towards the base of the cone; they fall off when ripe leaving a central axis on the tree.

Uses: soft yellow-white timber best for planks, joinery, boxes, carving, paper pulp, etc. Oil of turpentine, distilled from the leaves and wood, is used in medicine and veterinary work for sprains and bruises; Strasbourg turpentine, obtained from the bark blisters, is used in paints and varnishes.

mature female cone

GRAND or GIANT FIR
Abies grandis

A native of the west coast of North America, this fast-growing tree is planted for its timber in northern and central Europe, where it grows to 56m.
Crown: narrow and conical, with the branches in regular whorls.
Bark: brown-grey, with resin blisters, becoming darker and cracked into small square plates with age.
Shoots: smooth and olive green.
Buds: purple, becoming coated with resin; 2mm long.
Leaves: soft needles with notched tips, shiny green above with two silver bands beneath; 2–5cm long. They are arranged in two rows along the twigs.
Cones: erect and cylindrical, 7–10cm long, maturing from light green to red-brown.
Uses: pale cream or white timber for boxes, paper pulp, etc.

ATLAS CEDAR
Cedrus atlantica

Outside its native Atlas Mountains in North Africa, this cedar is commonly planted as an ornamental tree in parks and gardens and is sometimes grown for timber in southern Europe; it reaches a height of 40m.
Crown: broad and conical; the branches grow up from the trunk and their tips turn upwards.
Bark: smooth and dark grey, becoming cracked and scaly with age.
Buds: light red-brown and egg-shaped with black-tipped scales; 2–3mm long.
Leaves: stiff green or bluish-green needles, 1–3cm long, growing in tufts of up to 45 on short spurs; they form flat plates of foliage on the branches.
Male cones: conical, 3–5cm long.
Female cones: erect and barrel-shaped, with a hollow at the tip, maturing in 2 years to a pale purple-brown colour; 5–8cm long. The fan-shaped scales fall off to release winged seeds, leaving a central axis on the tree.

CEDAR OF LEBANON
Cedrus libani

A native of the Lebanon Mountains, Syria and south-east Turkey, the cedar of Lebanon is widely planted as an ornamental tree in parks, gardens, and churchyards; grows to 40m. It is distinguished from the Atlas cedar by the following features:
Crown: conical, becoming flat-topped with wide-spreading level branches.
Leaves: dark grass-green needles, 2–3cm long, growing in tufts of 10 to 20 on short spurs.
Female cones: like the Atlas cedar but larger – 7–12cm long.

NORWAY SPRUCE
Picea abies

A conifer of central and northern Europe, the Norway spruce is widely grown in forests, plantations, shelter belts, gardens, etc., for Christmas trees and timber; it grows to 40m.
Crown: narrow and conical; the branches are level except at the top, where they grow up from the trunk.
Bark: smooth and reddish-brown, becoming dark purple with age and flaking into small rounded scales.
Shoots: reddish or orange-brown.
Buds: smooth, brown, and pointed.
Leaves: dark-green sharp-pointed four-sided needles, 1–2cm long, growing from pegs all round the twigs.
Male cones: globular and yellow, 1cm long, hanging down from the tips of the shoots.
Female cones: erect, oval, green or dark red; when fertilized they become dark brown and cylindrical, 12–18cm long, and hang down.
Uses: strong light elastic pale-yellow timber for boxes, interior joinery, barrels, paper pulp, chipboard, violin and cello bellies, etc. The wood fibres are woven into mats and screens. Turpentine is extracted from blisters on the trunk and branches.

SITKA or SILVER SPRUCE
Picea sitchensis

Native to the coastal regions of western North America, from Alaska to North California, this fast-growing spruce is now widely planted in certain coastal areas of northern and western Europe for its useful timber; grows to 80m in North America, 50m in Europe. It can be recognized by the following features:

Crown: narrow and conical, with a long spire-like tip; in older trees the branches droop and the crown broadens.

Bark: dark grey or grey-brown and speckled; it flakes off in purplish scales in older trees.

Leaves: slender flat sharp-pointed needles, 1–3cm long, bluish-green above with two blue-white bands beneath.

Ripe female cones: light brown to whitish, blunt, and cylindrical, 5–10cm long; the scales are thin and papery with crinkled edges.

Uses: strong light fine-grained timber for interior joinery, boxes, carpentry, paper pulp, chipboard, pit-props, etc.

mature female cones

EUROPEAN LARCH
Larix decidua

Native to central Europe – from the Alps to the West Carpathians – this graceful deciduous conifer is planted in northern and western Europe as an ornamental tree and for its strong, highly-prized timber; it grows to 40m.
Crown: narrow and conical, becoming flat-topped with age. The branches tend to droop down, then turn up at the tips.
Bark: grey-brown and smooth, splitting into vertical cracks with age.
Shoots: pale yellow or pale pink, long, and hanging down from the branches.
Buds: brown, scaly, and resinous.
Leaves: soft needles, 2–3cm long, growing in clusters of 20 to 30 from short spurs on the twigs. Emerald green in March, they become darker, and finally golden in autumn.
Male cones: small, round, and golden, 0·5–1cm long.
Female cones: pale to rose-red, and flower-like, 1cm long, maturing to brown, egg-shaped cones, 2–4cm by 2–3cm, with rounded scales.
Uses: strong durable resinous timber, with yellowish sapwood and red-brown heartwood, used for fencing, gates, planking for fishing boats, staircases, wall panelling, light furniture, telegraph poles, railway sleepers, pit-props, etc.
The Japanese larch (*L. kaempferi*) is now widely planted for its quicker growth and greater disease resistance. Its mature cones have wavy edges and its needles are slightly wider than those of the common larch.

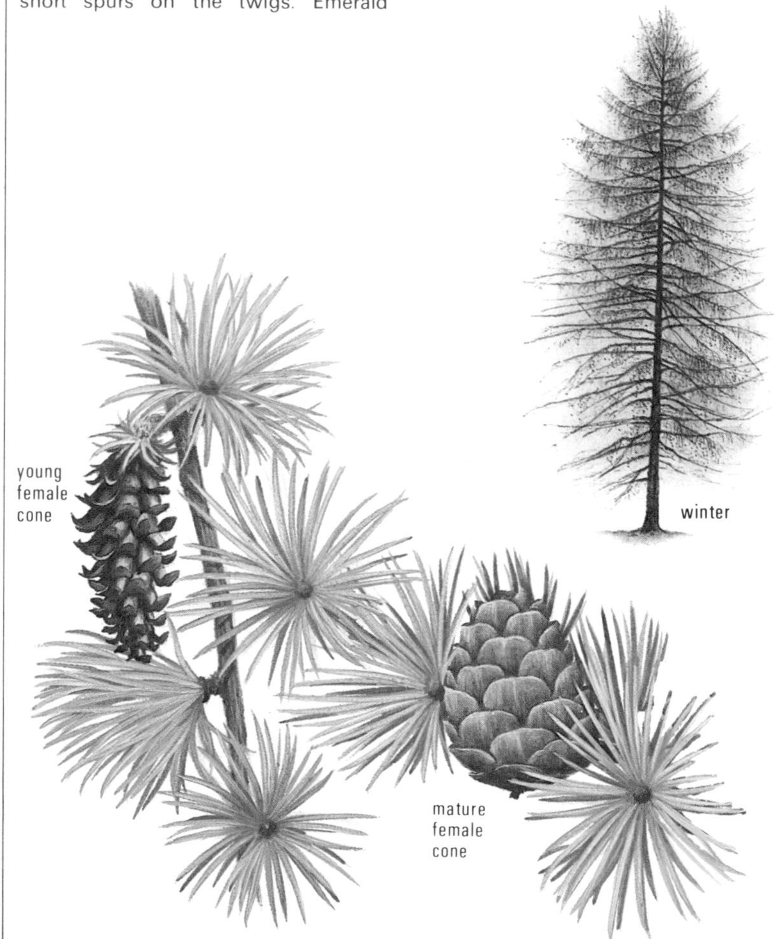

AROLLA or SWISS STONE PINE

Pinus cembra

This pine occurs at an altitude of just under 3000m in the Alps and Carpathians (which is close to the limit for tree growth). It is planted in other parts of Europe for ornament and timber and grows to a height of 25m.

Crown: columnar; stout level branches with upturned tips grow right from the base of the trunk.

Bark: dark grey or orange-brown, becoming scaly and rugged with broad cracks.

Shoots: greenish-brown, densely covered with orange-brown down.

Leaves: needles, 7–9cm long, growing in dense bunches of five; their outer surface is shiny dark green, the inner surface green-white.

Cones: egg-shaped, 8 by 6cm, ripening from purple to shiny red-brown. The edible seeds are released after the cone has fallen and rotted.

Uses: the pale resinous timber, which is easily worked, is used for furniture, turned articles, and toys.

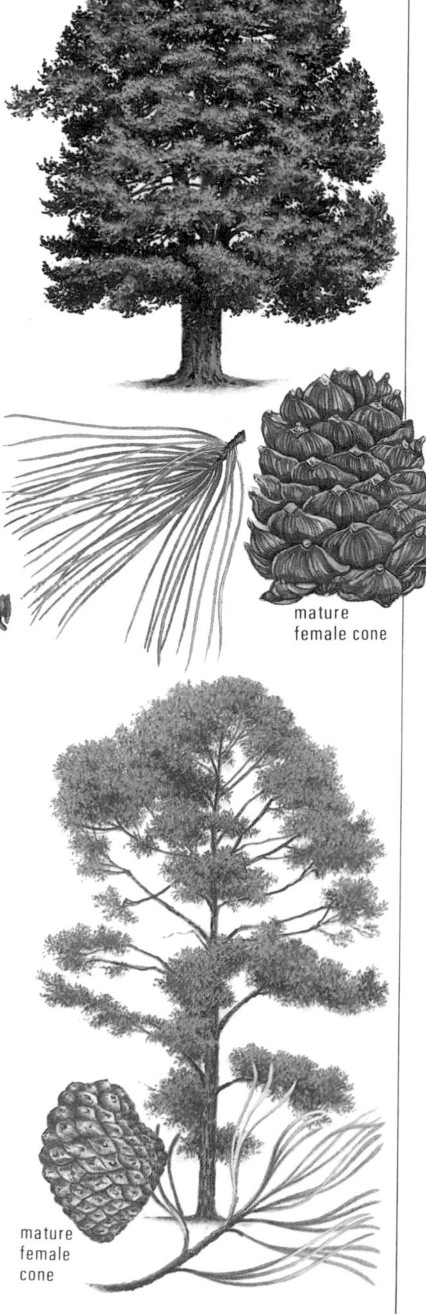

MONTEREY PINE

Pinus radiata

This wind-resistant tree, native to Monterey Bay in South California, is grown in mild coastal regions of western Europe for shelter and ornament; fast-growing, it is planted commercially in Spain, South Africa, Australia, and New Zealand. It grows to 30m.

Crown: conical, becoming broader and round-topped with age; branches grow low down on the trunk.

Bark: dark brown or dull grey, with deep cracks.

Buds: brownish, pointed, and covered with resin; 1–2cm long.

Leaves: slender bright-green needles, growing in threes; 10–15cm long.

Cones: large and irregular – 7–14cm by 5–8cm and flattened on the side nearest to the stem – with glossy brown scales. These cones grow in clusters of 3 to 5 and remain on the tree for many years.

Uses: timber for boxes and paper pulp.

WEYMOUTH PINE
Pinus strobus

A native of eastern North America, the Weymouth pine is widely cultivated in central and western Europe for its timber; it grows to 30–40m.
Crown: narrow and conical, becoming irregular and flat-topped with age.
Bark: smooth and greenish-brown in young trees, becoming grey-black and cracked with age.
Shoots: slender and bright green, becoming greenish-brown with fine hairs at the bases of the needle bundles.
Leaves: slender bluish-green needles, 8–12cm long, growing in bundles of five, form horizontal masses on the branches.
Cones: brown and banana-shaped, 10–15cm long, hanging down from the stem; scales curve outwards. Young cones are green and straighter.
Uses: pale-brown, light, fine-textured timber for pianos, stringed instruments, doors, window frames and general purposes.

AUSTRIAN PINE
Pinus nigra var. *nigra*

Native to Austria, central Italy, and the Balkans, this hardy wind-resistant pine is commonly planted on coasts as a wind-break and screen and to stabilize dunes; it grows to 33m.
Crown: irregular and spreading, with dark black-brown upper branches. Several stems normally rise from a short bole, making this a poor timber tree.
Bark: black-brown to dark grey, very scaly and coarsely ridged.
Shoots: shining yellow-brown, stout and ridged.
Buds: pale-brown and broad-based, 1cm long, tapering to a sharp point.
Leaves: stiff dark-green to black needles, curved and sharp-pointed, growing in pairs; 10–15cm long.
Cones: yellow- to grey-brown, egg-shaped and pointed; 5–8cm long.

mature
female cone

CORSICAN PINE
Pinus nigra var. *maritima*

This fast-growing pine, native to Corsica, southern Italy, and Sicily, is planted for shelter and yields a useful timber; grows to 35m. It can be distinguished from the Austrian pine by the following features:
Crown: narrow and conical, with level branches. One main trunk.
Bark: pink-grey to dark brown, with shallow cracks.
Buds: 2cm long, with a long point.
Leaves: flexible grey-green or sage-green needles, often twisted, growing in pairs; 12–18cm long.
Uses: hard strong timber, with ied-dish heartwood surrounded by pale-brown sapwood, used for general construction work in the Mediterranean; it also yields a useful resin.

STONE or UMBRELLA PINE
Pinus pinea

Native to the western Mediterranean, this wind-resistant pine has been planted in coastal regions all over the Mediterranean region since Roman times; it grows to 30m.
Crown: umbrella-shaped, with large spreading branches supported on a short trunk: an unmistakable feature of the Mediterranean landscape.
Bark: red-brown or orange, deeply cracked and scaly.
Shoots: pale greyish-green and curved.
Buds: red-brown, with deeply fringed white scales that turn out.
Leaves: dark greyish-green sharp-pointed needles, growing in pairs; 12-20cm long.
Cones: shining brown, globular, and flat-based, 10–15cm by 8–10cm; the scales are rounded and each has a central boss. The seeds are edible.
Uses: the rich oily seeds are eaten raw or roasted and used as a flavouring; the timber is used locally for furniture.

mature
female cone

SCOTS PINE
Pinus sylvestris

Native over much of Europe and northern and western Asia, the Scots pine is very widely planted for its valuable timber; it also makes an attractive hardy ornamental tree, thriving in light acid soils. Grows to 35–40m.

Crown: pyramid-shaped or conical when young, becoming flat-topped or rounded with age, with the branches sparsely arranged high up on the trunk.

Bark: at the base of the trunk reddish or grey-brown and cracked; on the upper trunk and branches orange-red to pink, and scaly.

Shoots: hairless and pale green, becoming brown.

Buds: cylindrical, dark brown or red.

Leaves: blue-grey or blue-green needles, often twisted, growing in pairs; 3–7cm long.

Male cones: small, yellow, and rounded, clustered near the tips of the shoots in early summer.

Female cones: pink and globular when fertilized, becoming green and turning down on the stem during the next year. Mature third-year cones, 3-8cm long, are grey-brown, ovoid. and pointed.

Uses: yields a good multi-purpose resinous timber with reddish heart-wood and pale-brown sapwood.

ALEPPO PINE
Pinus halepensis

A Mediterranean species, the Aleppo pine is a familiar sight in hot dry coastal regions. It thrives on exposed limestone hills and rocky ground, checking soil erosion and acting as a wind-break, and has many important uses in the Mediterranean region; it grows to 20m.

Crown: narrow when young, becoming domed with age with twisting branches supported on a stout trunk.

Bark: purple- or reddish-brown, with deep cracks.

Shoots: pale and slender, greenish-brown or yellowish.

Buds: red-brown, cylindrical, 1cm long.

Leaves: long slender flexible bright-green needles, usually in pairs; 9–15cm long.

Cones: bright reddish-brown, pointed or egg-shaped, 5–12cm long; borne on short down-turned stalks, they remain on the tree for several years.

MARITIME PINE
Pinus pinaster

The maritime pine is native to the coasts of central and western Mediterranean regions: it grows best in light well-drained soils and is often planted for shelter and to reclaim sand dunes, as well as for its timber; grows to 30m.
Crown: pyramid-shaped, with spreading branches growing high up on the trunk.
Bark: pale grey or reddish-brown with deep cracks, becoming darker with age.
Shoots: pinkish- or reddish-brown and hairless.
Buds: bright red-brown and non-resinous, with fringed scales.
Leaves: stout leathery sharp-pointed needles, growing in pairs; 15–25cm long.
Cones: bright glossy brown and pointed, up to 22cm long; the scales each have an upturned prickle. Cones remain on the branches for several years before opening.
Uses: tapped for resin, which yields turpentine oil.

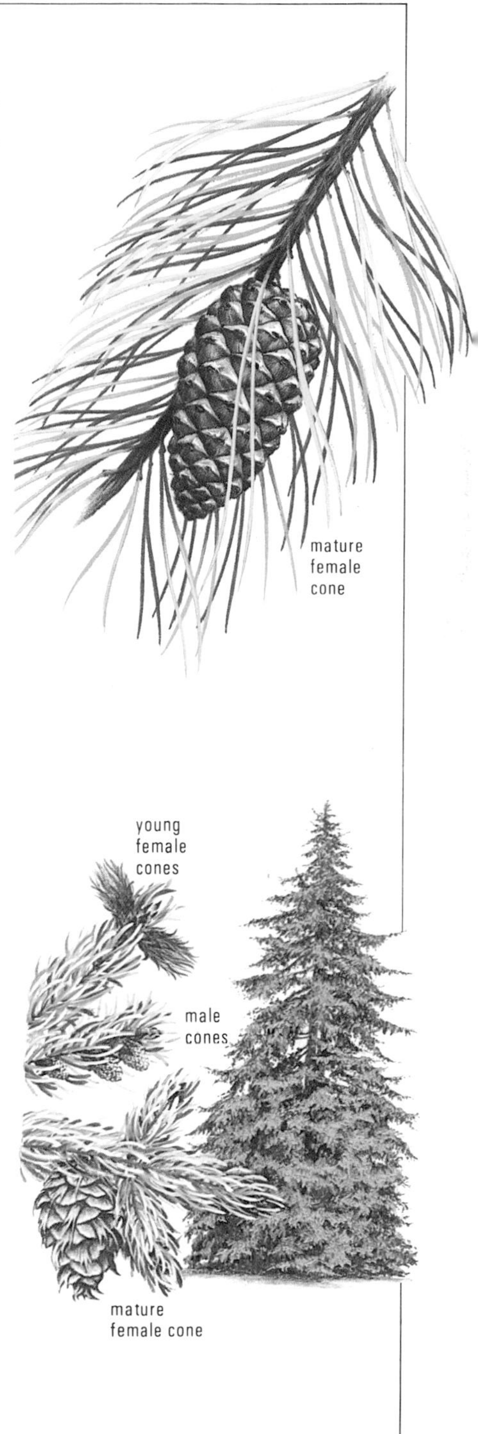

DOUGLAS FIR
Pseudotsuga menziesii

Native to western North America, this tall fast-growing tree is very widely planted in Europe for its timber. Growing to 100m in North America, it attains a height of 50m or more in Europe. It is not closely related to the true firs and can be distinguished from them by its cones.
Crown: pyramid-shaped, becoming flat-shaped with age.
Bark: dark grey and resin-blistered in young trees, becoming red-brown or purplish and corky, with deep cracks and ridges, with age.
Shoots: pale- or yellowish-green, covered with fine hairs.
Buds: pale brown and spindle shaped, up to 7mm long; non-resinous.
Leaves: flexible aromatic needles, 2–3 cm long, with two white bands underneath each side of the midrib; they grow singly and leave smooth oval scars when they fall.
Cones: dull brown and cylindrical 5–8 cm by 2·5cm, hanging down from the stem; the scales each have a three-lobed bract, which points towards the tip of the cone.

WESTERN HEMLOCK
Tsuga heterophylla

This graceful fast-growing conifer is native to the west coast of North America, where it may reach a height of 70m. A producer of good-quality timber, it is widely grown in north-west Europe; it prefers shade and is frequently planted under hardwood trees. It grows to 50m.
Crown: pyramid-shaped or conical, with a spire-like tip that arches over; the tips of the branches droop down.
Bark: brown, smooth, and flaky in young trees; becomes darker, with deep furrows and scaly ridges, with age.
Shoots: ribbed, pale yellow-brown, and covered with long hairs.
Buds: small, brown, and globular.
Leaves: flattened aromatic needles of different sizes 5–15mm long, with rounded tips; shining dark green above, they have two white bands along each side of the midrib on the lower surface. The needles have short stalks, and leave round orange scars on the twigs when they fall.
Cones: pale brown and egg-shaped, 2–3cm long, with smooth rounded scales; they hang down from the ends of the twigs. Unripe cones are green.

EASTERN HEMLOCK
Tsuga canadensis

A native of eastern North America, this hemlock is similar to the western species but its timber is of inferior quality. It is often planted in Europe for ornament and occasionally for timber; it grows to 32m. It can be distinguished from the other species by the following features:
Crown: broad and conical.
Buds: egg-shaped; green with a brown tip.
Cones: small – 1·5–2cm.

REDWOOD FAMILY Taxodiaceae

Most of the trees in this primitive family, which was once large and widely distributed, are now extinct. Those that remain – about 14 species – can be regarded as living fossils. They have awl-like or flattened needles and usually rounded, woody or leathery, cones. There are both evergreen and deciduous species.

COAST REDWOOD
Sequoia sempervirens

This majestic tree is native to western North America, its natural range being restricted to a narrow strip of coast from south-west Oregon to California. An important timber tree in North America, it is planted in Europe mainly for ornament in parks and gardens but occasionally for its timber. The redwood is considered to be the tallest tree in the world: a specimen of over 112m has been recorded in California; it grows to about 40m in Europe. It is also one of the longest lived: some Californian trees are over 2000 years old. It is an evergreen.

Crown: columnar, with level or drooping branches.

Bark: rusty red, soft, and fibrous, becoming darker, thicker, and deeply cracked with age.

Shoots: green and hairless, surrounded by green scale leaves.

Buds: short and scaly.

Leaves: hard, flattened, and blade-like, 1·5–2cm long, dark green above with a white band along each side of the midrib beneath; arranged in two rows along the side-branches. The leaves on the main stems of the branches are smaller and awl-shaped.

Male cones: small, rounded, and yellowish, in clusters at the tips of the main shoots.

Female cones: woody and globular, 2–2·5cm long, with wrinkled red-brown scales attached to the centre of the cone. They produce winged seeds.

GIANT SEQUOIA
Sequoiadendron giganteum

The most massive – but not the tallest – tree in the world, the giant sequoia forms natural forests in California's Sierra Nevada, where it grows to over 80m, with a girth of 24m, and lives for over 3000 years. In Europe it is planted for ornament, reaching a height of 50m and a girth of 7m.
Crown: narrow and conical; the ends of the branches curve upwards; the trunk is often buttressed at its base.
Bark: reddish-brown, thick, soft, and fibrous; becomes darker, fluted, and deeply cracked with age.
Leaves: blue-green or dark-green pointed scales, 4–7mm long, densely covering the branchlets. Evergreen.
Cones: brown and egg-shaped, 5–8cm long, drooping down from stalks at the ends of the branches; the scales are attached at the centre.

CYPRESS FAMILY Cupressaceae

The hundred or more species of evergreens in this family are found in the cooler regions of both hemispheres and on mountain tops in the tropics and subtropics. They often have two kinds of leaves – needle-like juvenile leaves and scale-like adult leaves – and small woody cones (except the junipers, which have fleshy cones).

MONTEREY CYPRESS
Cupressus macrocarpa

Native to California, the Monterey cypress – quick-growing and salt-resistant – is planted in western and southern Europe for both shelter and ornament in parks, gardens, churchyards and by the sea, and also for its timber. It reaches a height of 37m.
Crown: columnar, with a pointed top, becoming spreading and flat-topped with age.
Bark: brown with shallow ridges, becoming grey with thick peeling ridges in very old trees.
Leaves: scale-like and blunt-tipped, bright- or dark-green with paler margins; 1–2mm long. They completely cover the twigs and smell of lemon when crushed.
Male cones: yellow and egg-shaped, 3mm long, borne on small side-shoots.
Female cones: egg-shaped, green and purple, 6mm long, on central shoots; they ripen to rounded lumpy cones, 3–4cm long, with shining purple-brown scales each with a central boss.

LAWSON CYPRESS
Chamaecyparis lawsoniana

A tree of western North America – from north-west California and south-west Oregon — Lawson cypress is very widely planted in northern and central Europe for shelter and ornament. Many cultivated varieties are seen in parks, gardens, and churchyards; the tree is also planted commercially for timber on a small scale. Growing to 60m in America, it reaches a height of 38m in Europe.
Crown: narrow and conical; the tips of the branches droop down.
Bark: smooth, grey-brown, and shiny, becoming purplish-brown, cracked, and flaking off in older trees.
Leaves: bright green, triangular, and scale-like, borne on flattened branched twigs in horizontal sprays, like fern fronds.
Male cones: crimson and club-shaped, 5mm long, at the tips of the branches.
Female cones: green and globular, borne at the tips of the shorter branches, ripening to woody purple-brown cones, 7–8mm in diameter, producing winged seeds.
Uses: produces strong light durable timber, with yellow-white sapwood and dark-brown heartwood, for joinery, fencing, and underwater construction.

LEYLAND CYPRESS
× *Cupressocyparis leylandii*

This hybrid occurred in 1888 in the garden of C J Leyland as some chance seedlings, although their hybrid nature was not realized until the 1920s. The parents were the Nootka cypress (*Chamaecyparis nootkatensis*) and the Monterey cypress (*Cupressus macrocarpa*). Since then several hardy, fast-growing varieties have been raised and the tree is now widely planted in parks and gardens in north-west Europe. Growing to 30m, it is usually propagated by cuttings
Crown: columnar, tapering to a point.
Bark: dark red-brown, with shallow cracks.
Leaves: scale-like, of various colours (dark grey-green to blue-green, depending on the variety), clothing branched sprayed twigs.
Male cones: brown and club-shaped, yellow at the tips of the shoots.
Female cones: globular and greenish, ripening to grey or chocolate-brown cones, up to 1·5cm in diameter.

ITALIAN or FUNERAL CYPRESS
Cupressus sempervirens

This handsome Mediterranean cypress is commonly planted in gardens and cemeteries, especially in the Mediterranean region; it also yields a highly valued timber. The tree grows to 23m.
Crown: usually narrow and columnar, tapering to a pointed tip, but may be pyramid-shaped, with spreading level branches.
Bark: brown-grey, with shallow spiralled ridges.
Leaves: dark green, scale-like, and triangular, 1mm long, arranged in overlapping rows that completely cover the twigs.
Male cones: greenish and egg-shaped, 3mm long, at the tips of the twigs.
Female cones: greenish and globular, becoming dark red-brown and finally dull grey, 4 by 3cm; the scales each have a central spine.
Uses: strong durable fragrant timber, resistant to decay, used for carving, furniture, stakes, and vine props. The crushed leaves and seeds have medicinal properties.

COMMON JUNIPER
Juniperus communis

Very widely distributed in the northern hemisphere – from North America to south-west Asia and from Siberia to the Mediterranean – this adaptable small tree or shrub grows well on poor soils in a variety of habitats; it reaches a height of 6m.
Crown: variable – usually pointed but may be wide-spreading and broad.
Bark: reddish-brown.
Leaves: sharp-pointed needles, 1cm long, spreading out from the stems in whorls of three; there is a whitish band on the upper surface and the lower surface is grey-green.
Male cones: solitary, yellow, and cylindrical; 4mm long.
Female cones: greenish and globular, ripening in 2 to 3 years to blue-black berry-like fruits, 6–9mm in diameter.
Male and female cones grow on separate trees, in the axils of the needles.
Uses: the ripe berries are used for flavouring gin and seasoning food; oil of juniper is distilled from the unripe fruits.

CADE or PRICKLY JUNIPER
Juniperus oxycedrus

The prickly juniper is widespread in its native Mediterranean region, growing in coastal parts, dry hills, rocky ground, and woods; it reaches a height of 8m.
Crown: basically conical and densely branched. On windswept slopes it may be no more than a large 'cushion' hugging the ground.
Leaves: sharp-pointed or blunt needles, 16mm long, with two white bands along either side of the midrib on the upper surface.
Female cones: (on separate trees from the males): rounded and berry-like, ripening from green to reddish or yellowish; 6–10mm in diameter.
Uses: oil of cade, used in medicine and veterinary work, is distilled from the wood; the wood is very resistant to decay and is used for making charcoal.

WESTERN RED CEDAR
Thuja plicata

An important timber tree of western North America, the western red cedar is cultivated in parks, gardens, and plantations in Europe for shelter and timber. Noted for its quick growth, it reaches a height of 40m.
Crown: narrow and conical, with a spire-like tip and upswept branches; it broadens with age.
Bark: reddish-brown and fibrous, becoming grey-brown and peeling off in strips.
Leaves: blunt and scale-like, growing in sprays on flattened twigs; the upper surface is bright glossy green, the undersurface paler with white streaks.
Male cones: yellow and very small, at the tips of the smallest shoots.
Female cones: leathery and egg-shaped, ripening from green to brown, 1·5cm long; they are each made up of 10 to 12 thin overlapping spreading spine-tipped scales and are borne at the tips of the branches.
Uses: soft light durable timber, used for joinery, fencing and other outdoor construction.

YEW FAMILY Taxaceae

YEW
Taxus baccata

Widely distributed in Europe, North Africa, and south-west Asia, the yew is commonly planted in many cultivated varieties in parks, gardens, and churchyards. Yews can live to a great age – it is estimated that some are over a thousand years old – and reach a height of 25m; their bark, shoots, leaves, and seeds are all poisonous.
Crown: rounded or pyramid-shaped and densely branched; the branches are level or upturned.
Bark: reddish-brown and flaking; becomes deeply furrowed with age.
Leaves: leathery, sharp-pointed needles, 1–4cm by 3 mm, very dark green above and yellowish-green underneath. They are arranged in two rows along the side branches.
Male cones: small, rounded, and yellow, with overlapping scales.
Female cones: bright red and berry-like, enclosing a single seed; 1cm long.
Male and female cones grow on separate trees in the axils of the leaves.
Uses: the wood is hard, heavy, durable, strong, and elastic; in the Middle Ages it was used for bows, and more recently for cabinetwork, wood sculpture, etc.

MONKEY-PUZZLE FAMILY Araucariaceae

MONKEY PUZZLE
Araucaria araucana

The monkey puzzle – a native of Chile and Argentina – is quite widely planted in Europe as an ornamental tree for its curious branching system. It can grow to 30m but cultivated trees are smaller.
Crown: broad and rounded; the stout branches grow in regular tiers, drooping down at the base of the tree, and all have up-turned ends.
Bark: smooth and grey.
Leaves: leathery, green, and triangular, 3–4cm long, each with a spiny tip; they overlap each other and completely cover the twigs and branches.
Male cones: dark-brown and cylindrical, 10–12cm long.
Female cones: nearly spherical, 15cm across, green with golden spines; they break up into scales on the tree. The large brown seeds are edible and usually eaten roasted.
Male and female cones usually grow on separate trees.

spiky leaves

MAIDENHAIR FAMILY Ginkgoaceae

MAIDENHAIR TREE
Ginkgo biloba

The sole living member of a very ancient group, the maidenhair is a native of China. It is widely planted for ornament in parks and gardens. Although a gymnosperm, it has no cones and looks more like a flowering tree. It is deciduous and reaches heights of 30m. There are separate male and female trees.
Crown: variable, but normally tall and slender with short and elegant branches.
Bark: greyish brown, with a network of ridges and furrows when older.
Leaves: leathery, pale green, and fan-shaped – quite unlike those of any other tree. Up to 12cm across. Turn yellow before falling.
Male flowers: in thick yellow catkins 6–8cm long.
Female flowers: like small yellowish acorns: one or two on a slender stalk. Globular fruit 2·5cm across, green at first, ripening to brown.

leaves and male flowers

Flowering Trees

PALM FAMILY Palmae

This family of flowering plants is widespread in the tropics, and some palms are cultivated in subtropical and warm regions. Unlike the other trees, palms do not produce true wood, and their stems – which are unbranched – are covered with layers of strong tough fibres, rather than bark. The trunks do not get much thicker as they get older. Palm stems are crowned by a tuft of leaves. In the feather palms each leaf consists of a row of leaflets along each side of the midrib; in fan palms the leaflets arise from the same point at the top of the leaf stalk. The palms are all evergreens.

CHUSAN PALM
Trachycarpus excelsus

Native to south China and Japan, this tall fan palm is often grown as an ornamental both in the Mediterranean and also in warm regions elsewhere. The stout shaggy stem is covered with a mass of hard brown fibres and the woody bases of shed leaves; the tree reaches a height of 11m.
Leaves: rounded, each made up of 50–60 long narrow pointed leaflets. Borne on spiny leaf stalks, the leaves are a rich dark green, turning bright yellow and then dull brown before falling.
Flowers: yellow and very small, growing in drooping clusters, 60cm long; male and female flowers are usually in separate clusters.
Fruit: blue-black and globular; 1–1·5cm across.

DWARF FAN PALM
Chamaerops humilis

The only native European palm, the dwarf fan palm is found in dry regions along the Mediterranean coast – from Italy westwards but excluding France. It is widely cultivated; most forms are stemless but some have stout, fibre-covered trunks and may reach a height of 6–7m.
Leaves: rounded, each made up of 12 to 15 stiff pointed leaflets and borne on a slender spiny leaf stalk.
Flowers: small and yellow, in dense clusters that are at first sheathed in red spathes.
Fruit: brownish-yellow and globular; 2cm across.

WILLOW FAMILY Salicaceae

The willows and poplars constitute nearly all the 300 or more species of deciduous trees and shrubs that make up this family. They all have flowers grouped into single-sexed catkins. Willow catkins produce nectar and are pollinated partly by insects. Poplar catkins are always pollinated by the wind. The dry fruit – a capsule – splits to release seeds, which are covered in silky hairs.

WHITE POPLAR
Populus alba

Native to central and southern Europe and central and western Asia, the white poplar is widely planted as an ornamental in parks and gardens; it grows to 30m.
Crown: broadest at the top, with twisted spreading branches.
Bark: smooth and grey-white in young trees, becoming black and rough at the base and patchy above.
Shoots and buds: densely covered with white woolly down.
Leaves: either large (9 by 8cm) and 5-lobed or small (5 by 5cm) and usually oval, with toothed or lobed margins; all have stalks 3–4cm long. The undersurface and stalks are white and downy.
Male catkins: crimson and grey.
Female catkins: pale green or greenish yellow, producing fluffy seeds. Male and female flowers (catkins), 4–8cm long, grow on separate trees.

BLACK POPLAR
Populus nigra

Widely distributed over much of Europe, this poplar is often planted as an ornamental tree; it grows to 35m and there are many varieties and hybrids.
Crown: broad, with upturned branches; the trunk is short and thick and often carries large burrs.
Bark: grey-brown to black, deeply furrowed into broad ridges.
Leaves: triangular to diamond-shaped, 5–8cm by 6–8cm, with translucent toothed margins; borne on 3–4cm-stalks, they are deep green above and paler beneath, turning a soft yellow.
Male catkins: grey, becoming crimson; 5cm long.
Female catkins: greenish-white, 6–7cm long, producing white woolly seeds.
Male and female flowers grow on separate trees.
Uses: soft light nearly white wood used for packing cases and general purposes.

LOMBARDY POPLAR
Populus nigra var. *italica*

This variety of the black poplar, produced in Italy, is now very widely grown for ornament and for shelter-belts, where its quick growth is much appreciated: it grows to 36m.
Crown: narrow and columnar, with a pointed tip and upswept branches.
Leaves: triangular, with a rounded base and small curved teeth; 6 by 4·5cm. Bright green, they are borne on flattened stalks, 2·5cm long.

WEEPING WILLOW
Salix vitellina var. *pendula*

One of several types of weeping willow, this hybrid is widely planted as an ornamental tree in parks, gardens, and by rivers; it grows to 22m.
Crown: broad and domed; the curved branches bear long slender yellow shoots that hang straight down.
Bark: pale grey-brown, with a network of shallow ridges.
Leaves: narrow and pointed, 10cm long; pale green above, bluish-white beneath, and covered with fine hairs.

ASPEN
Populus tremula

Very widely distributed over the whole of Europe – and extending into the Arctic Circle – the aspen is most commonly found on hillsides and in damp places; it grows to 25m.
Crown: conical and sparsely branched, becoming broader with age.
Bark: greenish-grey and smooth, becoming brown and ridged at the base.
Buds: red-brown, sticky, and pointed.
Leaves: rounded, with curved irregular teeth; 4–6cm by 5–7cm. Borne on slender flattened stalks, 4–6cm long, they are dull rich green above and pale grey-green beneath. The leaves flutter in the slightest breeze, giving rise to the tree's alternative name of trembling poplar.
Male catkins: purplish-grey and downy, becoming yellow with pollen.
Female catkins: green, 4cm long, becoming woolly and white when they shed their seeds.
Male and female catkins grow on separate trees.
Uses: soft light white wood excellent for matches and paper pulp.

WHITE WILLOW
Salix alba

This willow is widely distributed in Europe, central Asia, and North Africa, often found growing by streams and rivers; it reaches a height of 25m.
Crown: conical, becoming rather shapeless with spreading branches.
Bark: dark grey with thick ridges.
Shoots: greyish-pink to olive-brown, slender, and hairy.
Buds: dark pinkish, covered with grey hairs.
Leaves: narrow and pointed, 7–8cm long, with toothed margins; blue-grey and covered with silky hairs.
Male catkins: yellow, with 2 stamens to each flower.
Female catkins: green and slender, becoming white and fluffy with seed.
Male and female flowers grow on separate trees.
Uses: light tough timber for flooring, cart bottoms, etc.; the pliant young twigs are used for basketry. Cricket bats are made from a quick-growing variety of this species known as the cricket-bat willow. It has purple shoots and bluer leaves than the normal white willow.

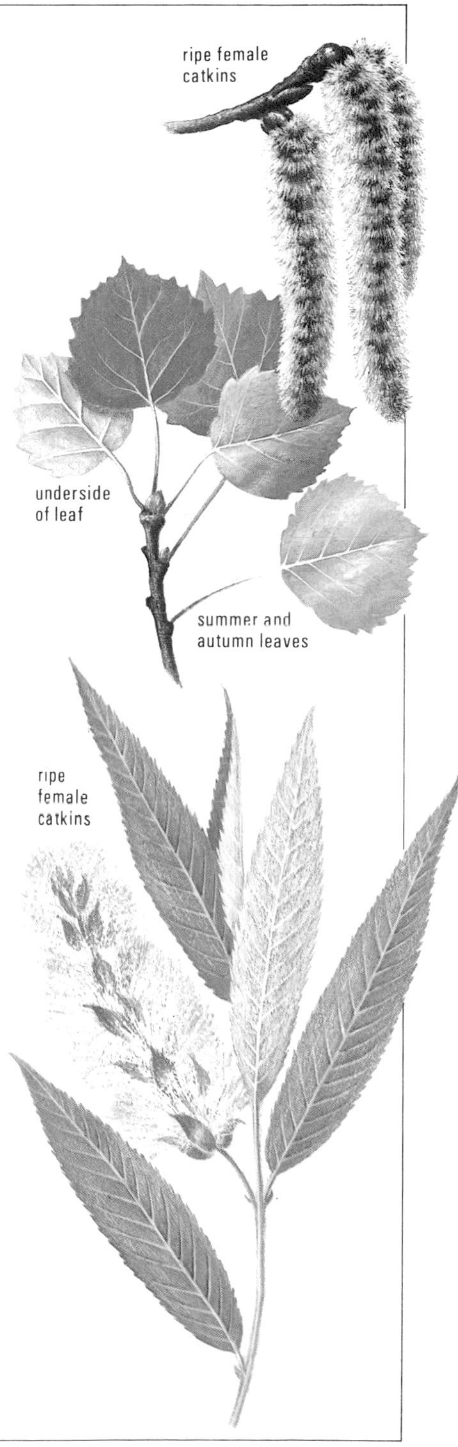

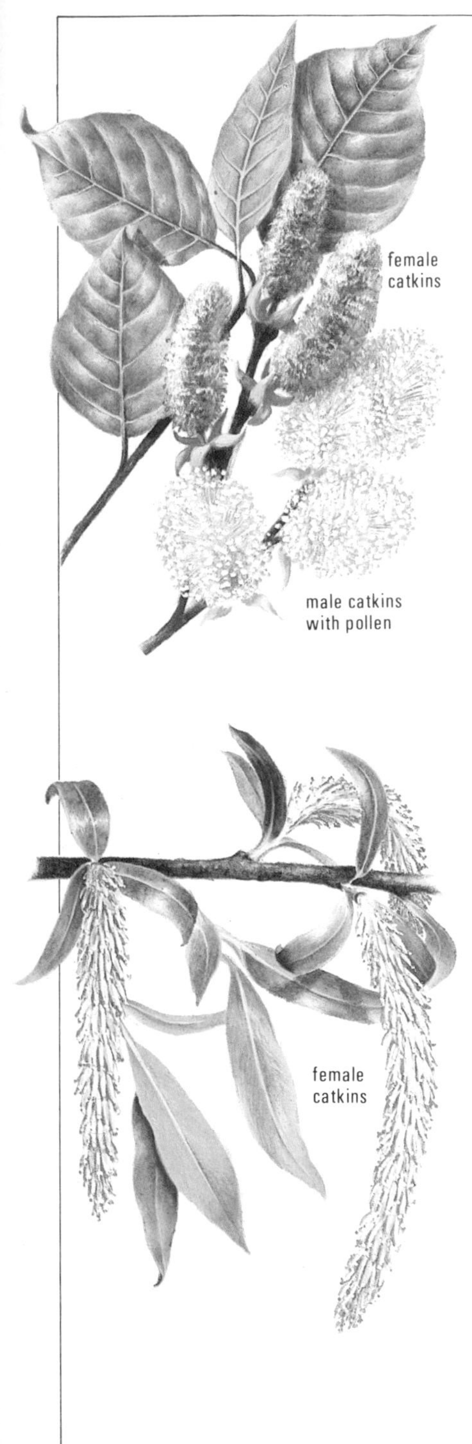

GOAT WILLOW or SALLOW
Salix caprea

Native from Europe to north-east Asia, the sallow is common in damp wooded regions and coppices; it grows to 16m.
Crown: open, with upswept branches.
Bark: smooth and grey in young trees, becoming brown with wide cracks.
Shoots: deep red-brown and initially covered with long hairs.
Buds: red, oval, and pointed; 3–4mm.
Leaves: usually oval with a pointed tip and wavy margins; 10 by 6cm. Dark grey-green above and grey and woolly beneath, they have dark red hairy stalks with 2 small leaves at the base.
Male catkins: egg-shaped, 3cm long, and covered with silvery hairs; later they sprout golden-tipped stamens.
Female catkins: arched, slender (5–6cm long), and pale green with whitish styles, producing fluffy seeds. Male and female flowers grow on separate trees and appear long before the leaves.

CRACK WILLOW
Salix fragilis

Found all over Europe and as far east as western Siberia and Iran, the crack willow is common in damp places, e.g. by rivers; it grows to 25m.
Crown: broad and conical with upswept branches, becoming domed with twisted branches.
Bark: grey and scaly, developing thick brownish ridges with age.
Shoots: greenish-brown, snapping off cleanly and readily at the base.
Buds: brown, slender, and pointed.
Leaves: narrow and pointed, 12cm long, bright green and glossy above, grey-green and waxy beneath.
Male catkins: yellow; 2–5cm long.
Female catkins: green, 10cm long, becoming white and fluffy with seed. Male and female flowers grow on separate trees and appear with the leaves; they are slender, cylindrical, and pointed.

BIRCH FAMILY Betulaceae

The deciduous trees and shrubs of this family – about 120 species – grow in northern temperate regions. Their flowers are grouped into separate male and female inflorescences on the same tree: the males are in long drooping catkins; the females in shorter catkins or erect clusters. Hornbeam and Hazel are now commonly placed in a separate family.

HORNBEAM
Carpinus betulus

The natural range of this slow-growing wind-resistant tree is from the Pyrenees to southern Sweden and east as far as south-west Asia. It makes excellent hedges and produces hard timber; it grows to 30m.
Crown: rounded, with upswept branches and a deeply fluted trunk.
Bark: smooth and pale grey, sometimes with fine pale-brown stripes.
Buds: pale brown, slender, pointed, and turned in towards the stem.
Leaves: oval and pointed, with reddish stalks, double-toothed margins, and about 15 pairs of prominent parallel veins; 8–10 by 5–6·5cm. Very dark green above and yellowish beneath, they turn old gold in autumn.
Male flowers: bright yellow-green drooping catkins, 2·5–5cm long.
Female flowers: shorter catkins, made up of green leafy bracts each carrying two crimson-styled flowers. They develop into clusters of 3-lobed bracts, each 3·5cm long and bearing a pair of small nutlets.
Uses: tough heavy nearly white wood used for chopping blocks, mallets, skittles, wooden rollers, etc.

DOWNY BIRCH
Betula pubescens

This slender and elegant tree resembles the silver birch but prefers damper soils. It is common in fens and on wet heaths and it forms extensive stands in the Scottish Highlands. It completely replaces the silver birch in some northern areas. Its bark never has the black diamonds of the silver birch. The branchlets do not droop and are clothed with soft hairs. The petiole is also densely hairy, while the leaf is more rounded than that of the silver birch and its margins are uniformly toothed.

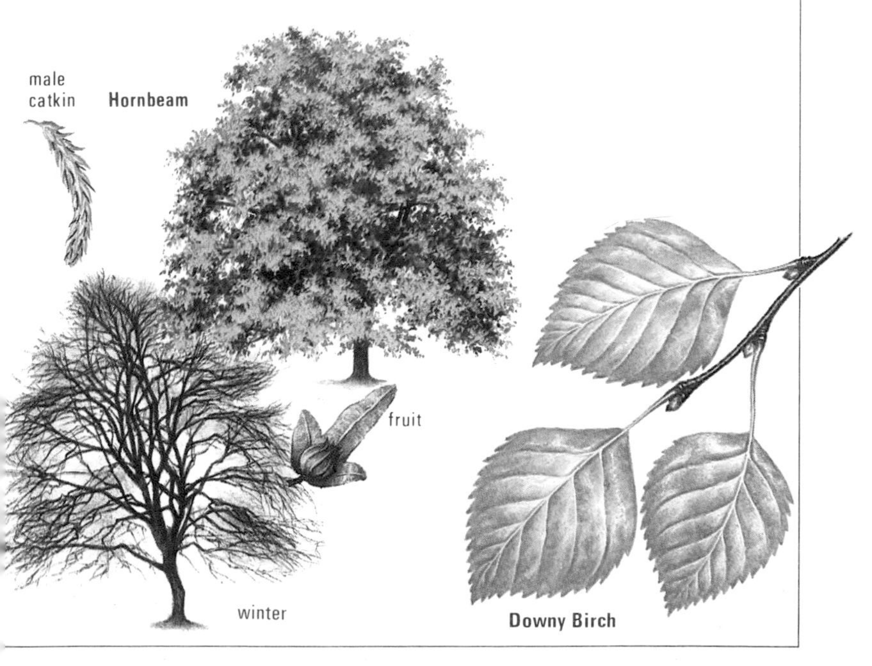

SILVER BIRCH
Betula pendula

This graceful slender tree, native to most of Europe and south-west Asia, grows well on light peaty sandy soils; it reaches a height of 30m.
Crown: narrow and conical with upswept branches, becoming rounded, with long hanging branchlets and a deeply fluted trunk.
Bark: shiny purplish-brown in young trees, becoming pinkish-white and finally white with black diamond-shaped markings; smooth and peeling above, black and knobbly at the base.
Shoots: dark purple-brown, with raised white warts.
Leaves: emerald green and triangular, with rounded bases and double-toothed margins; 3–7cm long.
Male flowers: clusters of 2 to 4 drooping yellow catkins, 3cm long, at the tips of the shoots; young catkins are pale purple-brown and visible all winter.
Female flowers: clusters of about 6 catkins on branched stalks below the males; at first erect, green, and club-shaped, 1–1·5cm long, they become brown and hang down, 2–3cm long, and release small winged fruits.
Uses: the hard strong pale-brown wood is used for small turned articles and, in Scandinavia, for plywood, flooring, and skis; the twigs are used for brooms and brushes and the bark for roofing, tanning, etc.

COMMON ALDER
Alnus glutinosa

Found all over Europe and also in Siberia and North Africa, the common alder grows by open water – from mountain streams to lowland fens; it reaches a height of 25m.
Crown: broad and conical or pyramid-shaped; the spreading branches are at first upswept and later level.
Bark: purplish-brown, becoming dark grey-brown and cracked into small square plates.
Shoots: green and sticky, becoming

purple with orange markings.
Buds: green to purple, 7mm long, borne on short stalks, 3mm long.
Leaves: oval, with a pointed base, a rounded tip, and wavy or toothed margins; 10 by 7cm. Pale orange-brown when they first open, they become very dark green.
Male and female flowers (catkins) appear before the leaves.
Male catkins: in clusters of 3 to 5, maturing from dull purple to dark yellow; 5cm long.
Female catkins: present all the year round in short erect clusters; dark red catkins, 5–6mm long, mature into green egg-shaped cones, 8–15mm long, which become dark brown and woody when ripe.
Uses: the wood is strong, easily worked, and durable under water; it is used for piles, barrels, toys, broom-heads, hat blocks, etc., and paper pulp. This alder is often planted to conserve river and lake banks; it also improves the fertility of the soil.
The grey alder (*A. incana*) is similar, but can be distinguished by its duller leaves which are sharply pointed and strongly toothed. It flourishes in dry soils and is often planted when rubbish tips and similar places are being reclaimed.

HAZEL
Corylus avellana

Widely distributed in Europe and south-west Asia, the hazel is found in woods, thickets, and hedgerows; it can grow to 12m but is usually shorter (6m).
Crown: usually a broad bush, sometimes with a short trunk.
Bark: smooth and shiny grey-brown, with horizontal rows of pores.
Shoots: pale-brown, covered with long swollen-tipped (glandular) hairs.
Buds: smooth, blunt, and egg-shaped, changing from brown to green.
Leaves: rounded with a pointed tip and double-toothed margins, up to 10 by 10cm, borne on hairy stalks. Hairy and deep green above, they turn brown and finally yellow.
Male flowers: brownish-yellow catkins that appear in autumn, becoming yellow and longer (5cm) by spring.
Female flowers and fruit: small brown buds with protruding crimson stigmas develop into clusters of 1 to 4 nuts, each partly enclosed in a toothed green husk; the nuts change from whitish-green to pale pink-brown and finally brown by autumn.

WALNUT FAMILY Juglandaceae

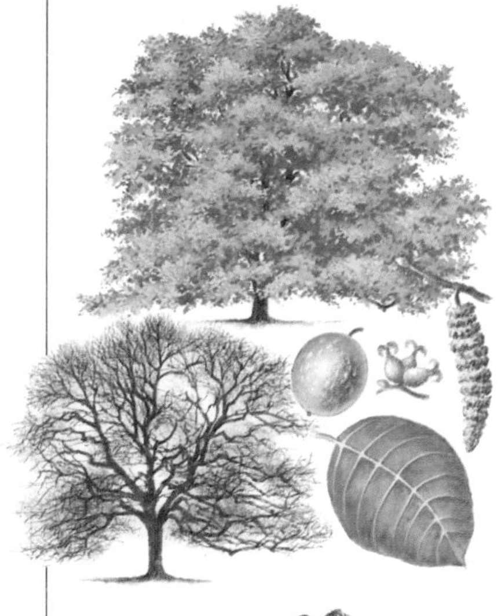

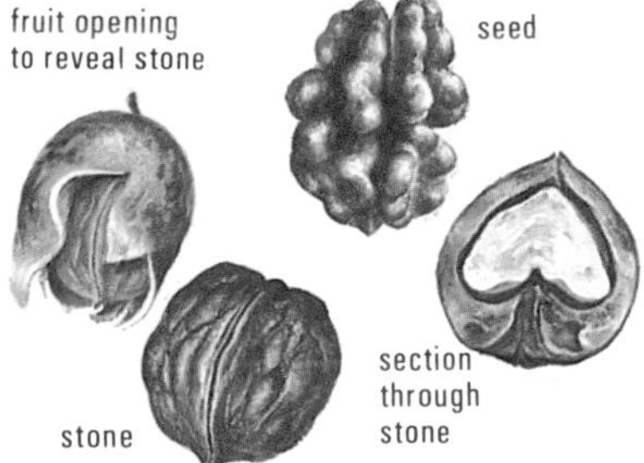

WALNUT
Juglans regia

Thought to be native to south-east Europe and south-west Asia, the walnut is now naturalized in many parts of Europe; it grows to 30m.
Crown: rounded or spreading.
Bark: very pale grey, becoming deeply furrowed with age.
Buds: broad and squat; deep purple-brown to black.
Leaves: compound, with 3 to 4 leaflets down each side of the stalk and one (the largest) at the tip. Each leaflet is leathery and oval with a pointed tip, 8–20 by 4–10cm, and aromatic when crushed. Orange-brown when they first open, they become dull green above and paler beneath.
Male flowers: greenish-purple catkins, 5–10cm long, in the leaf axils.
Female flowers: greenish-yellow, 1cm long, in erect clusters of 2 to 5 at the tips of the shoots. They develop into globular dark-green fruits (drupes), the outer layer decaying to reveal the stone within.
Uses: the seeds are edible and yield a useful oil; the unripe fruits are pickled. A fast brown dye is obtained from the fruits and other parts. Walnut timber is hard, heavy, and fine-grained, with a pale grey-brown sapwood and chocolate-brown heartwood, and is attractively figured; it is highly valued for furniture, gunstocks, etc.

BEECH FAMILY Fagaceae

This large and commercially important family – containing about 1,000 species of deciduous or evergreen trees and shrubs – is widely distributed in all temperate regions of the world. The male flowers are borne in catkins, while the female ones are borne singly or in small clusters. The fruit is a nut, partly or completely enclosed in a cup-like husk.

BEECH
Fagus sylvatica

Native to most of Europe (except northern Scandinavia), the beech is a dominant forest tree; it is also widely planted for ornament, shelter, hedges, and timber. It grows to 30m.
Crown: slender and conical, becoming rounded with spreading branches.
Bark: smooth and silvery-grey.
Buds: slender and pointed, 2cm long, covered with brown papery scales.
Leaves: oval with a pointed tip, wavy margins, and 5 to 7 parallel veins on each side; 10 by 7cm. Clear green and silky at first, they become dark shiny green above and paler beneath with hairs on the larger veins, turning pale yellow and finally rich orange-brown before falling.
Male flowers: rounded greenish-yellow clusters, each of about 15 tiny flowers, on long drooping stalks.

Female flowers: in a rounded green head on a short stiff hairy stalk.
Male and female flowers grow in separate clusters on the same tree and open with the leaves.
Fruit: a pointed green husk, 2·5cm long, covered with soft green hairs; turns brown and splits into 4 to release 1 to 2 shiny brown nuts, triangular in section.
Uses: the strong, hard, fine-grained wood, bright buff with brown flecks, is used for furniture and turnery (e.g. tool handles, bowls, spoons, chair legs); the nuts provide mast for pigs, cattle, and poultry.

winter bud

ripe fruit in husk

CORK OAK
Quercus suber

Native to south Europe and North Africa, the evergreen cork oak is planted for ornament in parks and gardens as well as commercially (mainly in Portugal and south-west Spain) for its corky bark; grows to 20m.
Crown: domed and spreading, with low heavy twisting branches.
Bark: very rugged, with thick spongy ridges of pale brown or pale grey cork between wide dark cracks. Stripped trunks are pinkish-red.
Leaves: oval and pointed, with 5–6 shallow spine-tipped lobes on each side; 4–7 by 2–3cm. Blackish-green above and densely hairy beneath, they have hairy stalks, 1cm long.
Fruit: acorns, 1·5–3cm long, in deep cups with spreading upper scales.
Uses: cork, removed from the trunk every 8–10 years, is used for bungs, shoe soles, flooring, floats, life buoys, etc. The heavy wood is used for joinery and fuel.

HOLM OAK
Quercus ilex

This evergreen south European oak is widely planted for shelter and ornament, especially by the sea, being resistant to salty winds and pollution in towns; it grows to 30m.
Crown: dense and rounded, usually with a short trunk and straight upgrowing branches.
Bark: brownish-black to black, cracked into small square plates.
Shoots: slender, dull greyish-brown, and woolly.
Buds: fawn and downy, 1–2mm long.
Leaves: vary from long and narrow to oval, with spiny-toothed, wavy, or smooth margins; 5–10 by 3–8cm. The upper surface is rough and shiny blackish-green, the lower surface greyish-green and densely hairy; the leaf-stalks are woolly and 1–2cm long.
Male flowers: in pale-gold catkins, 4–7cm long.
Female flowers: grey-green and hairy with pink tips, 2mm long, growing on woolly stalks in clusters of 2 to 3.
Fruit: light-green acorns, 1·5–2cm long, with deep cups covered with rows of grey-haired fawn scales.
Uses: the hard, heavy, tough highly figured wood is used for wheels, joinery, vine-props, fuel, and charcoal; the bark is used for tanning leather and dyeing.

SESSILE or DURMAST OAK
Quercus petraea

The sessile oak – native to Europe (including Britain) and west Asia – forms forests over much of its natural range. It grows best on light acid soils and reaches a height of 30–40m.
Crown: open and domed, with straight branches radiating from a straight trunk.
Bark: grey, with fine, usually vertical, cracks and ridges.
Leaves: oblong, with a wedge-shaped base and 5 to 9 pairs of rounded lobes; 8–12 by 4–5cm. Dark green and leathery, they are borne on long yellow stalks (1–2cm).
Male flowers: in slender pale-green catkins.
Female flowers: tiny and greenish-white, with red-purple stigmas.
Fruit: rounded acorns, either stalkless or on very short stalks (5–10mm), ripening from green to brown.
Uses: (timber) see Pedunculate oak.

PEDUNCULATE or COMMON OAK
Quercus robur

The most widespread European oak: a long-lived slow-growing tree occurring in forests, woods, parks, and gardens all over Europe, from Spain to North Africa, north-east Russia, and south-west Asia; it grows to 45m. It can be distinguished from the sessile oak by the following features.

Crown: wide and domed, with wide-spreading branches (the lower ones are massive and twisted).

Leaves: oblong, with an ear-like lobe at the base on each side of the stalk and 4–5 pairs of rounded lobes with wavy or toothed margins; 10–12 by 7–8cm. Borne on short stalks (4–10mm), they are dull dark green above and paler beneath, turning rich orange-brown in autumn.

Fruit: acorns, 1·5–4cm long, with shallow cups, usually growing in pairs on stalks 4–8cm long. They mature from whitish-green to dark brown.

Uses: strong heavy timber, with white sapwood and golden-brown heartwood, is durable and resistant; used for furniture, fencing, gates, railway carriages, panelling, chests, etc., and, in the past, for shipbuilding. The bark is used in tanning leather and the acorns provide mast for pigs.

Pedunculate Oak

RED OAK
Quercus borealis

This oak from eastern North America is commonly planted in Europe as an ornamental tree, for its attractive autumn foliage; for shelter; and, particularly in central Europe, for timber. It grows to 35m.

Crown: broad and domed, with straight radiating branches and a short straight trunk.

Bark: smooth and silvery-grey.

Leaves: oblong, 12–22cm long, with a pointed base and tip and 4 to 5 sharply angled lobes on each side, the tip of

each lobe extending into a bristle. Borne on yellow stalks, 2–5cm long, the leaves turn from pale yellow to dark green above, pale grey beneath, and become dull red or red-brown in autumn before they fall.
Fruit: flat-based dark red-brown acorns, 2 by 2cm, in shallow scaly cups with incurved rims and stout 1-cm stalks.

TURKEY OAK
Quercus cerris

The Turkey oak, native to south-west Asia and south and central Europe, is now widely naturalized in Europe. One of the fastest-growing oaks, it is widely planted for shade and ornament; it grows to 38m.
Crown: wide and broadly domed with straight up-growing branches.
Bark: dull dark-grey and roughly cracked.
Buds: pale brown and downy, surrounded by long twisted whiskers.
Leaves: variable, but usually with 7 to 14 deep triangular lobes down each side; 9–12 by 3–5cm. Borne on hairy stalks, 2cm long, the leaves are rough and dull green, becoming shiny, above; paler and woolly beneath.
Male flowers: in catkins, 5–6cm long, maturing from red to yellow.
Female flowers: oval, 5mm long, with dark-red stigmas surrounded by yellowish scales.
Fruit: narrow egg-shaped acorns, 2·5 by 1·4cm, either stalkless or on very short stalks and surrounded by a mossy cup, 1cm deep, with long pointed pale-green scales.

SWEET or SPANISH CHESTNUT
Castanea sativa

This Mediterranean tree is widely grown for its edible nuts – it should not be confused with the horse chestnut whose seeds (conkers) are inedible. Long-lived and fast-growing, it does best on dry sandy soils; it reaches a height of over 30m.

Crown: conical and open when young, becoming columnar, and finally rounded and spreading.

Bark: silvery-grey, but becoming dark with deep, spirally arranged, cracks.

Shoots: stout and shiny purple-brown; smooth or downy.

Buds: rounded, yellow-green to red-brown.

Leaves: oblong, with a pointed tip and prominent parallel veins each extending into a bristly tooth on the margin; 10–25 by 9–10cm. Borne on red or yellowish stalks, 2·5cm long, they turn from bronze to glossy dark green, and finally pale yellow or rich brown in autumn.

Male flowers: minute with long stamens, growing in clusters on yellow catkins 10-12cm long. Catkins appear in summer.

Female flowers: in groups of 1–3 at the base of late catkins of male flowers. Each is surrounded by a green spiny cup from which the styles protrude.

Fruit: shiny red-brown nuts grouped in pairs or threes in a yellow-green husk, 3 by 4cm, covered with radiating spines. The nuts are released when the husk splits.

Uses: the nuts are eaten roasted and used to make flour, bread, etc., and for fattening livestock. The timber, grown as a coppice crop, is much used for fencing.

HOLLY OAK
Quercus coccifera

Also known as the kermes oak, this evergreen species is abundant on the hot, dry hillsides of the Mediterranean basin. It can reach 6m in height, but more commonly grows as a dense holly-like bush just a metre or so high. On some hillsides these oaks form impenetrable thickets.

Leaves: bronze and slightly hairy when young, becoming dark green and very prickly. 2–4cm long and hairless below when mature.

Flowers and Fruit: male catkins yellow, 2–4cm long, appearing with young leaves in spring. Acorns, hidden among foliage, ripen in second summer: cup clothed with sharp spines.

unripe acorn in spiny cup

ELM FAMILY Ulmaceae

SMOOTH-LEAVED ELM
Ulmus carpinifolia

Native to Europe, North Africa, and south-west Asia, this species is the common elm of continental Europe; it grows to 30m.
Crown: tall, narrow, and domed, with the branches growing up nearly vertically from the trunk and arching over into long hanging branchlets.
Bark: grey-brown, with long deep vertical cracks and long thick ridges; bark on the branches has fine black vertical cracks.
Shoots: pale brown, slender, and hairless.
Buds: egg-shaped, red, and hairy.
Leaves: oval, 6–8cm long, with a pointed tip, an oblique base, and toothed margins. Borne on hairy stalks 5mm long, the leaves are bright shiny green above.
Flowers and fruit: small red flowers with white stigmas appear before the leaves and develop into transparent winged fruits.

ENGLISH ELM
Ulmus procera

This elm is native to Britain and occurs in many varieties and local forms in southern and central Europe; it grows to 35m.
Crown: tall, narrow, and domed, with massive twisting branches growing upwards from high up on the trunk.
Bark: dark brown or grey, deeply cracked into small square plates.
Shoots: long, slender, reddish-brown, and densely hairy.
Buds: pointed, dark brown, and downy; 2–3mm.
Leaves: rounded, or oval and pointed, with double-toothed margins, an oblique base (one side may be lobed), and 10 to 12 pairs of veins; 4–10 by 3·5–7cm. Borne on 5mm downy stalks, they are dark green and rough with hairs on the upper surface and turn yellow or bright golden in autumn.
Flowers and fruit: dark purplish-red flowers with tufts of stamens appear before the leaves. They develop into winged fruits, with the seed close to the notched tip of the membrane.
Uses: the reddish-brown timber is strong, firm, heavy, and does not split easily; used for coffins, indoor and outdoor furniture, and – since it is durable under water – for bridges, piles, and groynes. The inner bark has medicinal properties.

WYCH ELM
Ulmus glabra

Native to northern and central Europe and west Asia, the wych elm grows naturally in woods and hedgerows and is often planted in exposed situations and polluted atmospheres; it grows to 40m.

Crown: broadly domed, with branches spreading and arching out from low down the trunk, which is often forked.

Bark: smooth and silvery-grey in young trees, becoming brown with a network of broad grey-brown ridges.

Shoots: stout and dark red-brown, covered with hairs when young.

Buds: pointed and dull red-brown with a covering of reddish hairs.

Leaves: oval, with a pointed tip, double-toothed margins and a very unequal base (the base on one side forms a rounded lobe that covers the stalk); 10–18 by 6–9cm. Borne on thick hairy stalks, 2–5mm long, the leaves are dark green and very rough above, paler with soft hairs beneath.

Flowers and fruit: dark purplish-red flowers grow in dense clusters that appear before the leaves. They develop into round winged fruits, 2–5cm wide with the seed at the centre of the membrane, which ripen from light green to brown.

Uses: the pliable strong wood is used for boat- and carriage-building, tool handles, shafts, and furniture.

NETTLE TREE
Celtis australis

A graceful deciduous tree from the Mediterranean region, often planted for ornament and shade in the south. It reaches heights of 25m.

Crown: an irregular dome with flexible, drooping shoots.

Bark: greyish brown, smooth with just a few gentle wrinkles.

Leaves: elongated heart-shaped with prominent teeth and wavy margins and often with a long, twisted point. Rough above and softly hairy below. 10-15cm long.

Flowers: small and green, hanging on long stalks among the leaves.

Fruit: cherry-like, 9-12mm in diameter: green at first, becoming dark brown or black when ripe. The flesh is edible, with a sweet taste. The fruits of related American trees are known as hackberries.

MULBERRY FAMILY Moraceae

FIG
Ficus carica

Native to west Asia, the fig is widely cultivated – both as an important fruit crop (mainly in south Europe) and for shelter and ornament (it is often trained against walls); it grows to about 10m.
Crown: spreading, with stout knobbly upswept branches.
Bark: smooth and metallic grey, finely patterned in darker grey.
Leaves: thick and leathery, up to 30 by 25cm, with a heart-shaped base and 3 to 5 coarsely toothed lobes, the middle lobe being the largest. Borne on stalks 5–10cm long, they are dark green and rough with hairs above and beneath.
Flowers: tiny, enclosed in a fleshy pear-shaped structure with a small hole at the top through which pollinating insects enter.
Fruit: dark green and pear-shaped, becoming larger and either violet or blackish when ripe.
Uses: the fruit is eaten either fresh or dried; it also has laxative properties.

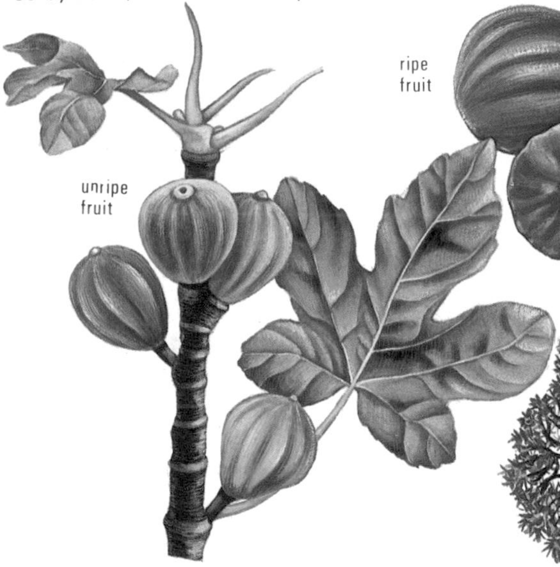

ripe fruit

unripe fruit

BLACK or COMMON MULBERRY
Morus nigra

The black mulberry, native to central and west Asia, is widely cultivated in south Europe for its fruit and is grown elsewhere mainly for ornament; it reaches a height of 12m.
Crown: low and broadly domed, with rough stout twisting branches arising from a short trunk.
Bark: dark orange, with wide cracks and many bosses and burrs.
Shoots: stout and downy, turning from pale green to brown.
Buds: stout and pointed, shiny dark purplish-brown.

Leaves: heart-shaped, with toothed or lobed margins; 8–12 by 6–8cm. Growing on stout hairy stalks, 1·5–2·5cm long, they are rough, hairy, and deep green above, paler and finely hairy beneath.
Flowers: male and female flowers grow in separate pale-coloured catkins, the males being short and stout and the females rounded.
Fruit: rounded and raspberry-like, made up of a cluster of fleshy blobs each surrounding a central seed. Green at first, they become orange-scarlet and finally deep blackish-red and sweet enough to eat.

PLANE FAMILY Platanaceae

LONDON PLANE
Platanus × hispanica

A fast-growing hybrid between the Oriental plane and the American plane (*Platanus occidentalis*), this tree is widely planted for shade and ornament in city streets and squares; it is resistant to pollution, thrives in restricted rooting space, and withstands heavy pruning. Grows to 35m.
Crown: domed, with large spreading branches supported on a long trunk.
Bark: smooth, thin, and grey-brown, flaking off to reveal greenish or yellow patches.
Shoots: pale green, becoming stout and brown.
Buds: conical and red-brown, with a large protruding base and covered with a single scale.
Leaves: 5-lobed, each lobe being triangular with coarsely toothed margins. Borne on red-brown tube-like stalks, the leaves are bright shiny green above and paler beneath.
Flowers: male and female flowers grow in separate rounded clusters hanging on long stalks on the same tree. The males are yellow; the females crimson.
Fruit: in brown globular clusters, 3cm across, that remain on the tree all winter and break up the following spring. Each fruit is 1cm long with a style projecting from the top and a parachute of yellow hairs at the bottom.

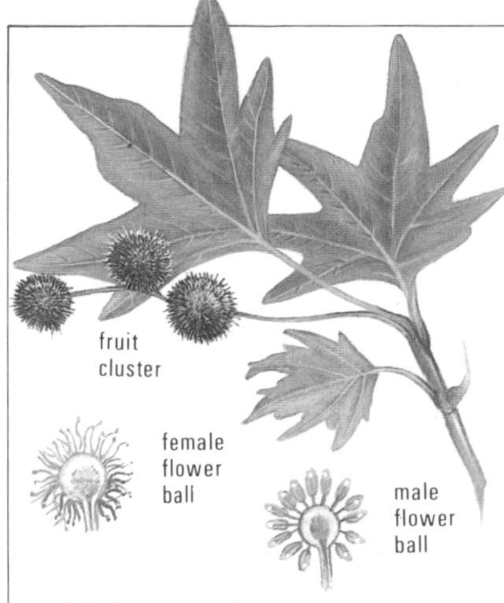

ORIENTAL PLANE
Platanus orientalis

Native to south-eastern and eastern Europe, Asia Minor, and India, the Oriental plane is often planted for shade and ornament in southern and eastern Europe; grows to 30m. Slower growing than the London plane and less tolerant of pollution, it can be distinguished from this tree by its leaves.

Leaves: 5- to 7-lobed, 18 by 8cm; the lobes are longer and narrower than those of the London plane. Borne on yellowish stalks, 5cm long, each with a thick red base, the leaves turn from pale orange-brown to yellow-green and finally to pale bronze-purple.

LAUREL FAMILY Lauraceae

SWEET BAY or BAY LAUREL
Laurus nobilis

In its native Mediterranean region this attractive evergreen tree reaches a height of 20m; it is widely grown elsewhere as an ornamental pot plant or shrub. Laurel leaves – worn in wreathes as a sign of victory or honour in classical times – are today used in cooking to season food.

Crown: dense and conical, with spreading up-growing branches.

Bark: smooth and blackish, with paler cracks in older trees.

Leaves: lance-shaped with wavy margins and dark-red basal veins; 5–10 by 2·5–3cm. Borne on dark-red stalks, 6mm long, they are leathery and very dark green above, yellow-green beneath, and aromatic when crushed.

Flowers and fruit: pale yellow inconspicuous flowers, 1cm across, develop into shiny berries, 8–10mm across, ripening from green to black.

female flowers

male flowers

ROSE FAMILY Rosaceae

Distributed all over the world, this vast family contains over 2,000 species of trees, shrubs and herbaceous plants, including many important fruit trees. The members are distinguished by their flowers, which have 4-5 petals and an equal number of sepals.

CRAB APPLE
Malus sylvestris

The crab apple commonly grows in woods, thickets, and hedgerows of Europe and south-west Asia; it reaches a height of 10m.
Crown: dense, low, and domed, with many twisting spiny branches.
Bark: greyish-brown or dark brown, splitting into small thin square plates.
Shoots: ribbed and often thorny; dark purple above, pale brown beneath.
Buds: small (4–5mm) and pointed, dark purple and covered with grey hairs.
Leaves: oval, with a rounded or wedge-shaped base, pointed tip, and toothed margins; 5–6 by 3–4cm.
Flowers: small, with 5 white petals, usually tinged with pink, and many yellow stamens.
Fruit: globular, 2·5 by 2·8cm, with a hollow at each end and a central 'core' containing brown seeds (pips). The apples are glossy yellow-green with white spots and become speckled or flushed with red in autumn.
Uses: the fruit, though too sour to be eaten, is made into crab-apple jelly; the red-brown wood – hard, tough, and fine-grained – is used for ornamental carving, mallet handles.

MEDITERRANEAN MEDLAR
Crataegus azarolus

Widely distributed in the Mediterranean region, this shrub or small tree is cultivated for its fruit in southern Europe; grows to 4-12m.
Shoots: downy.
Leaves: pale green in colour, 3–7cm long, divided into lobes usually without teeth and with hairs on the lower surface.
Flowers: small and white, with purple-tipped stamens, growing in downy clusters 5–7·5cm across.
Fruit: large (2–2·5cm across) and round, orange-red or yellow, and containing 1 to 3 nutlets.
Uses: the fruit, which has a pleasant slightly acid taste, is made into jams.

MEDLAR
Mespilus germanica

Native to south-eastern Europe and western and central Asia, the medlar has long been cultivated, especially in western and central Europe, for its fruit; it reaches a height of 6m.
Crown: low and spreading, with tangled branches.
Bark: grey-brown and deeply cracked into oblong plates that flake off.
Shoots: downy; they sometimes develop spines.
Leaves: oblong, with a pointed tip and smooth or toothed margins; 15 by 5cm. The upper surface is dull green, with indented veins, and is sometimes hairy; the lower surface is paler and densely hairy. The leaves are borne on very short (5mm) hairy stalks.
Flowers: stalkless, 3–6cm across, with 5 broad white petals, 5 long (4cm) green sepals, and many brown-tipped stamens.
Fruit: globular, 5–6cm across, ripening from green to brown. Persistent sepals surround an open pit at the tip through which the brown 'seeds' (actually individual fruits) can be seen.
Uses: the fruit is edible only when soft and over-ripe; it can also be made into a jelly.

ripe fruit

ripe fruit

MIDLAND HAWTHORN
Crataegus laevigata

Similar to the next species, but much less common and found mainly on heavy soils – especially in damp woodlands. It can be distinguished from the common hawthorn by the following features:
Leaves: have shorter and blunter lobes, with the incisions rarely reaching half way to the mid-rib; lobes always toothed; no hair tufts on lower surface.
Flowers: usually with 2 styles.
Fruit: normally with two persistent styles at tip and two nutlets.
Hybrids between the two species are very common. Both species have cultivated varieties with pink or red flowers – the red hawthorns. Some of these have double flowers.

HAWTHORN or MAY
Crataegus monogyna

This small spiny tree is very widely distributed in Europe, growing in thickets, hedgerows, and at the edges of woods; it is also planted as a windbreak and boundary hedge. It reaches a height of 10m.
Crown: spreading or rounded, with intertwining branches.
Bark: smooth and brown at first, becoming darker and rugged.
Shoots: dark purple-red or reddish-brown with straight thorns, 1–2·5cm long.
Buds: very small, reddish-black, and scaly.
Leaves: divided into 3 to 7 deep lobes with smooth or sparsely toothed margins; up to 8 by 7cm (usually 3·5 by 4cm). Shiny green above: tufts of hairs at the bases of the veins beneath.
Flowers: 8–15mm across, with 5 white overlapping petals, purple-tipped stamens, and one style; they grow in dense fragrant clusters of 16 or more.
Fruit: round, 8–10mm across, with a persistent style at the tip and containing (usually) one nutlet; ripens from green to dark red.
Uses: the heavy dense wood has been used for tool handles, mallet heads, and other small articles; makes good charcoal.

BIRD CHERRY
Prunus padus

Native to northern and central Europe and Asia Minor, the bird cherry grows in woods, especially by streams, as a low shrub or a tree up to 15m tall.
Crown: rounded, with sharply ascending upper branches and spreading or drooping lower ones.
Bark: smooth and dark brown, with a strong unpleasant smell of bitter almonds.
Shoots: olive green, turning dark brown.
Buds: slender and sharply pointed; shiny brown.
Leaves: oval, with a pointed base and tip and finely-toothed margins; 10 by 7cm. Borne on dark red stalks, 2cm long, they are dull green above and pale green beneath, turning pale yellow or red in autumn.
Flowers: small, white, and fragrant, grouped in long (8–15cm) spreading or drooping clusters.
Fruit: globular and shiny, 8mm across, ripening from green to black. Bitter-tasting, they are dispersed by birds.

WILD CHERRY or GEAN
Prunus avium

In its wild state this tree grows in woodlands in most parts of Europe; it reaches a height of 20m. It is the ancestor of all cultivated forms of sweet cherry and is widely grown in many varieties both for its fruit and for its blossom and attractive autumn foliage.

Crown: spreading, with branches growing up from a tall straight trunk.
Bark: purplish-grey, marked with horizontal orange-brown corky ridges and peeling off in thin horizontal strips.
Shoots: greyish-brown.
Buds: shiny red-brown and pointed.
Leaves: oval, with a pointed tip, finely-toothed margins, and 2 glands near the base; 10 by 4·5–7cm. Drooping from stalks 2–3·5cm long, which are red above and yellow beneath, the leaves are pale green (downy beneath) and turn crimson or yellow in autumn.
Flowers: white and sweetly scented, growing on slender stalks in clusters at the tips of the branches.
Fruit: rounded and shiny, 2·5cm across, growing on a red-brown stalk 3–5cm long and ripening from green to bright red and finally purple. Sweet-tasting when ripe, they are dispersed by birds.
Uses: the fruit of cultivated varieties is eaten fresh, made into jams, liqueurs, etc.; the golden-brown wood is heavy, hard, and tough and prized for furniture and turned articles (such as bowls, pipes and similar musical instruments). Large trunks yield valuable veneers, with a gleaming surface and interesting grain patterns.

MYROBALAN or CHERRY PLUM
Prunus cerasifera

Native to the Balkans and central Asia, the myrobalan is planted in central Europe for its edible fruit (the cultivated plum is probably a hybrid between this tree and the blackthorn); it grows to about 8m. There are also several early-flowering cultivated varieties, which are widely grown as ornamentals.

Crown: open and spreading.
Bark: brownish-black.
Shoots: smooth and glossy green.
Leaves: oval with blunt-toothed margins; 4–7cm long. Borne on purple-green stalks, 1cm long, the leaves are glossy green above, paler and matte beneath (some ornamental varieties have reddish leaves).
Flowers: white, 2cm across, with 5 petals; cultivated ornamental varieties have white or pink flowers.
Fruit: globular and grooved down one side, ripening from pale glossy green to yellow or red, and containing a flattish stone.

CHERRY LAUREL
Prunus laurocerasus

A native of S. E. Europe and western Asia, this large evergreen is commonly grown for ornament and hedging, and also as cover for pheasant rearing. It is naturalised in many woodlands. Reaching 14m in some western areas, it is more often a spreading bush, casting such deep shade that nothing grows beneath it. The plant is often simply called laurel, but this causes confusion with the bay laurel or true laurel (page 184). The leaves contain prussic acid (cyanide): entomologists sometimes crush them and use them for killing insects.

Crown: usually broad and spreading.
Bark: brownish grey with prominent lenticels (breathing pores), often a squared pattern.
Leaves: thick and leathery, oval, up to 20cm long: bright and shiny on upper surface, pale green at first and becoming darker with age.
Flowers: creamy white and fragrant, carried in erect spikes in spring. Plant flowers only when growing freely in good light: clipped bushes rarely flower.
Fruit: purplish black berries up to 2cm in diameter.

ALMOND
Prunus dulcis

A native of west Asia and North Africa, this small tree is widely cultivated – in warm regions for its seeds, and elsewhere for ornament in gardens. It reaches a height of about 6m.

Crown: open and rounded, with up-growing branches.
Bark: purplish-black, deeply cracked into small square plates.
Leaves: oval, 7–12cm long, with a pointed tip and finely toothed margins. Dark green or yellowish green in colour, they are often folded along the midrib into a V shape.
Flowers: pink, 3–5cm across, with 5 petals and many stamens; they open well before the leaves. Some cultivated varieties have white or double flowers.
Fruit: oval and pale green, 4cm long, splitting when ripe to reveal a pale brown stone within which is the edible kernel or seed.
Uses: seeds from the sweet almond are eaten raw and used for cooking, flavouring etc. The hard reddish wood is used for veneers.
The peach (*P. persica*) is very similar but has a large juicy fruit.

BLACKTHORN or SLOE
Prunus spinosa

The blackthorn is widely distributed in Europe and parts of Asia, growing in hedgerows, on waste ground, scrub, hillsides, etc.; it produces suckers and grows to a height of 4m.
Crown: dense and upright, with a tangled mass of thorny branches.
Bark: black; in old trees it is deeply cracked into small square plates.
Buds: small, oval, and pointed; reddish-purple to black in colour.
Leaves: small (4cm long), dull green, and oval, with a bluntly pointed tip and shallow-toothed margins.
Flowers: white, 1–1·5cm across, with 5 petals and long orange-tipped stamens; they usually open well before the leaves.
Fruit: globular, 1·5–2cm across, ripening from green to purple-black with a waxy bloom; the green flesh has a very bitter taste.
Uses: the fruit is used for jams and jellies, flavouring gin, and is fermented into sloe wine.

APRICOT
Prunus armeniaca

A native of Asia, the apricot is widely grown in S. Europe for its fruit. Susceptible to frost, it is grown further north only in sheltered places.
Crown: sturdy and rounded, to 10m.
Bark: brownish grey, often ridged.
Leaves: oval, to 10cm long, with long reddish stalks: reddish when young, becoming bright green.
Flowers: white or pale pink in early spring, before leaves appear.
Fruit: yellow to orange, with furry coat: stone smooth.

ROWAN or MOUNTAIN ASH
Sorbus aucuparia

This attractive tree grows wild in woodlands and rocky mountainous regions of Europe, south-west Asia, and North Africa; it is also widely planted as an ornamental tree in streets, parks, and gardens. It reaches a height of 20m.
Crown: oval and open, with up-growing branches.
Bark: smooth and shiny silver-grey, becoming light grey-brown and marked with a network of thin scaly ridges.
Shoots: purplish- or brownish-grey, hairy at first, becoming smooth.
Buds: dark purple-brown, 1·7cm long, covered with grey hairs.
Leaves: compound, 22cm long, consisting of 5 to 7 pairs of leaflets and one terminal leaflet (each is oval, with toothed margins; 6 by 2cm). Hairy at first, they become smooth and deep green above, grey-green beneath; the leaves of some cultivated varieties turn bright red in autumn.
Flowers: creamy white and 5-petalled, 1cm across, growing on woolly stems in flat-topped sweetly scented clusters, 10–15cm across.
Fruit: round berries, 1cm across, maturing from yellow to orange and finally scarlet.
Uses: the fruit, rich in vitamin C, is made into jelly; the smooth hard purple brown wood is used for carved and turned articles.

SNOWY MESPILUS
Amelanchier ovalis

This small, much branched shrub grows on rocky hillsides in southern and central Europe, especially on limestones. The young leaves and twigs are clothed with white fur, and when the flowers open in spring the bush appears completely white – hence its common name. It reaches 4m in height.
Bark: dark grey to black on mature branches, but pale grey and furry when young.
Leaves: oval and coarsely toothed, 1·5cm long: clothed with white hairs below when young, soon becoming hairless and bright green.
Flowers: 2cm across, white with narrow, widely separated petals: in small clusters.
Fruit: small and round, blue-black with sweet, edible flesh.

WHITEBEAM
Sorbus aria

The whitebeam, native to southern and central Europe and parts of Britain, is found in woodlands, rocky regions, and on southern mountains; reaches a height of 25m. It grows well on chalk and limestone, and, because it withstands pollution, it is often planted in city streets.
Crown: domed, with up-swept radiating branches.
Bark: smooth and grey, developing shallow cracks and ridges with age.
Shoots: brown and hairy at first, becoming smooth and grey.
Buds: green, with brown-tipped scales and a white hairy tip; 2cm long.
Leaves: oval, with shallow-toothed or lobed margins; 8 by 5cm. The upper surface is dull green, the lower surface is densely covered with white hairs, giving the whole tree a glistening white appearance when the leaves first appear. The leaves turn yellow or pale brown in autumn and finally pale grey before falling.
Flowers: white, 1·5cm across, growing on white woolly stalks in clusters 5–8cm across.
Fruit: rounded, 8–15cm across, ripening from green to orange-red and finally deep scarlet; they are dispersed by birds.
Uses: the hard, heavy, tough wood – yellowish-white and fine-grained – is sometimes used for handles, spoons, etc.; the fruit is made into a jelly.

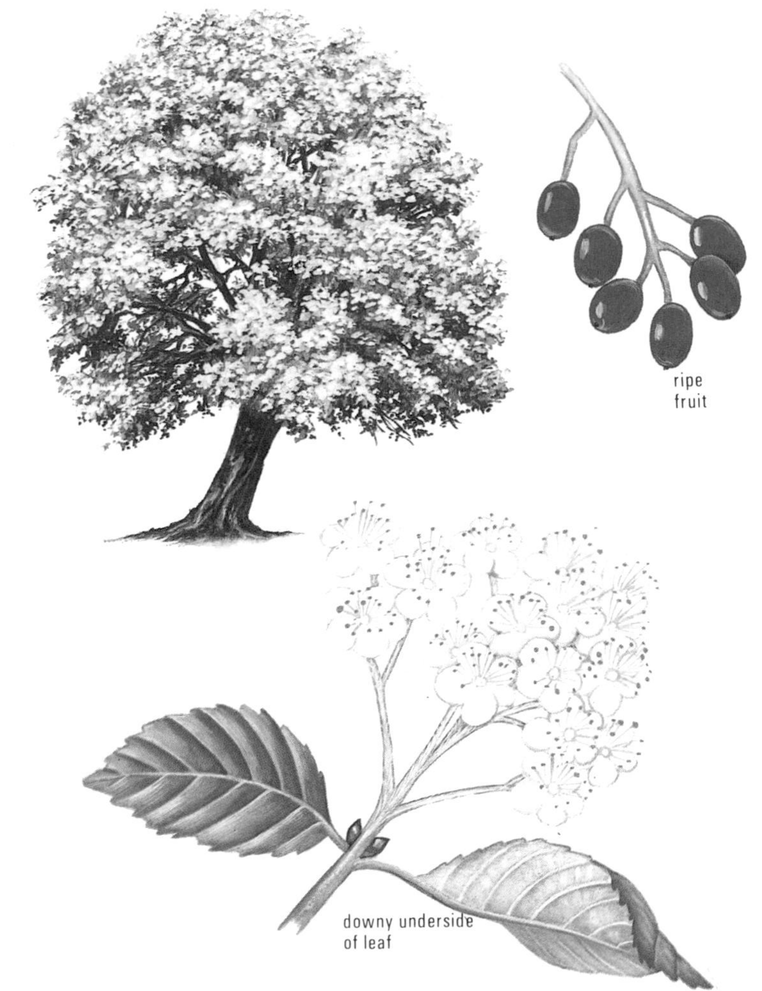

WILD SERVICE TREE
Sorbus torminalis

This tree is widely distributed in Europe (except the north), North Africa, and parts of Asia; it grows to a height of 25m.
Crown: conical when young, becoming domed and spreading, with up-growing branches.
Bark: pale grey to black-brown; it cracks into thin plates that flake off.
Shoots: brown and shiny.
Buds: glossy green and globular; 4–5mm long.
Leaves: divided into 3 to 5 pairs of triangular toothed lobes that decrease in size towards the tip of the leaf; 10 by 8cm. Borne on yellowish-green stalks, 2–5cm long, the leaves are shiny deep green above and yellow-green beneath, turning yellow, deep red, and purple in autumn.
Flowers: 1·2cm across, with 5 white petals and yellow stamens; they are grouped into loose domed clusters, 10–12cm across.
Fruit: oval, 1cm or more long, ripening from green to brown with rust-coloured specks. Acid-tasting, they are said to have medicinal properties and have been used in the past as a cure for colic.

BASTARD SERVICE TREE
Sorbus × thuringiaca

A hybrid between the rowan (mountain ash) and the whitebeam, the bastard service tree is grown for ornament, mostly in town streets; it reaches a height of 15m.
Crown: oval and upright, becoming dense and leaning to one side.
Bark: dull grey with shallow cracks.
Shoots: pink-grey with a purple tip.
Buds: dark red-brown; 8mm long and with few scales.
Leaves: oblong, 11 by 7cm, with lobes that decrease in size towards the tip of the leaf; 1–4 pairs of toothed leaflets grow at the base of each leaf. Borne on stout red stalks, 2cm long, the leaves are dark grey-green above and white with down beneath.
Flowers: white, 1cm across, grouped in downy clusters, 6–10cm across.
Fruit: bright red, 1·2cm across, growing in clusters of 10 to 15.
The Swedish whitebeam (*S. intermedia*) is a very similar tree, but its leaves lack free leaflets at the base. A native of the Baltic region, it is widely planted for ornament in towns. It is tolerant of smoke pollution.

TRUE SERVICE TREE
Sorbus domestica

The service tree – native to southern Europe, North Africa, and west Asia – is widely planted for ornament and (particularly in central Europe) for its fruit; it grows to a height of 20m. When not bearing fruit, it can be distinguished from the rowan – a similar species – by its bark and buds.
Crown: domed, with spreading level branches.
Bark: orange- to dark-brown, cracked (often deeply) into rectangular plates.
Buds: egg-shaped, glossy bright green, and resinous; 1cm long.
Leaves: compound, 15–22cm long, with 6 to 10 pairs of leaflets and one terminal leaflet, each oblong and sharply toothed, 3–6cm long, dark yellow-green above, downy beneath.
Flowers: 1·5–2cm across, with 5 cream-coloured petals, triangular sepals, and 5 styles; they grow in domed erect clusters, 10 by 14cm.
Fruit: globular or pear-shaped, 2–3cm long, ripening from green to brown. They are edible when over-ripe and are used in continental Europe for making alcoholic beverages.

PEAR
Pyrus communis

The parent species from which the numerous orchard and garden varieties of pear are derived, this tree grows wild in woods, hedgerows, etc., of Europe and west Asia; it reaches a height of 20m.
Crown: slender, with a rounded top and up-growing branches.
Bark: grey-brown or black, breaking into small deep squares.
Shoots: brown, often downy, and sometimes thorny.
Leaves: oval to heart-shaped, 5–8cm long, with a pointed tip and smooth or toothed margins. Borne on long (2–5cm) stalks, they are glossy dark- or yellow-green.
Flowers: 2–4cm across, with 5 white petals and dark-red stamens; they grow in dense clusters, 5–8cm across, that open before the leaves.
Fruit: varies from globular to oblong, 2–4cm long, ripening from green to brown; the flesh is sweet when ripe.
Uses: the hard compact pinkish-brown wood is used for furniture, turned articles, wood blocks, etc.; it also makes good fuel and charcoal.

PEA FAMILY Leguminosae

The 7000 or more species of this family – which includes peas, beans, and other herbaceous plants as well as trees and shrubs – are found all over the world. They are distinguished by their fruit – a pod. Because their roots bear nodules containing nitrogen-fixing bacteria, leguminous plants improve the fertility of the soil in which they grow.

FALSE ACACIA or LOCUST TREE
Robinia pseudoacacia

Native to eastern North America, the false acacia has long been planted in Europe as an ornamental tree in parks, gardens, etc. It grows well on sandy soils and is often planted to stabilize the soil; it reaches a height of 30m.
Crown: irregular and open, with twisted branches and a fluted and burred trunk.
Bark: smooth and brown in young trees, becoming dull grey and rugged, with a network of deep ridges and cracks, with age.
Shoots: dark red and ribbed, each with a pair of short spines at its base.
Buds: small and hidden by the leaf-stalks until autumn.
Leaves: compound, 15–20cm long, made up of 3 to 7 pairs of oval leaflets and one terminal leaflet (each 2·5–4·5cm long). The leaves vary from yellow-green to light green and spines are often present at the base of the leafstalks.
Flowers: white and resembling those of the pea, growing in dense hanging fragrant clusters, 10–20cm long.
Fruit: dark brown pods, 5–10cm long, each containing 4 to 16 seeds; they hang from the branches in bunches well into winter.

LABURNUM
Laburnum anagyroides

The common laburnum grows wild in woods and thickets in mountainous regions of southern and central Europe; it reaches a height of 7m. It is very widely planted as an ornamental tree in parks, gardens and streets, the hybrid *L. × vossii* being particularly popular. All parts of the tree, especially the seeds, are poisonous.

Crown: narrow, open, and irregular, with up-growing branches.
Bark: smooth; green at first, becoming greenish-brown.
Shoots: grey-green and covered with long grey hairs.
Buds: egg-shaped, pale grey-brown, and hairy.
Leaves: compound, with 3 oval pointed leaflets, 3–8cm long. Borne on stalks 2–6cm long, the leaves are greyish-green above, blue-grey and covered with silky hairs beneath.
Flowers: bright yellow and shaped like those of the pea, 2cm long, growing in hanging clusters, 10–30cm long.
Fruit: slender pods, 4–8cm long, containing black seeds. Hanging in bunches, they are hairy when young, becoming dark brown and hairless.

JUDAS TREE
Cercis siliquastrum

Said to be the tree on which Judas Iscariot hanged himself, the beautiful Judas tree grows wild in rocky regions of southern Europe and western Asia. It is often planted – especially in warmer regions – for ornament in parks and gardens; it reaches a height of 10–12m.

Crown: low and irregularly domed.
Bark: purplish, becoming pinkish-grey with fine brown cracks.
Shoots: dark red-brown.
Buds: dark red, narrow, and conical, 3–5mm long.
Leaves: nearly circular, with a heart-shaped base and smooth margins; 8–12 by 10–12cm. Borne on stalks 5cm long, they are yellow- or dark-green above and paler beneath.
Flowers: rosy pink and resembling those of the pea, 2cm long, growing in clusters (often directly from the trunk) that open before the leaves have appeared.
Fruit: flat red-purple pods, becoming brown and remaining on the tree throughout winter.

GORSE
Ulex europaeus

A spiny evergreen shrub which covers large areas of heath and rough grassland. Its golden, scented flowers are favoured by bees. On being touched, the flowers explode pollen on to the visiting insect. Can reach 3m when growing erect, but in exposed places or when subject to heavy grazing it forms low cushions.

Leaves: trifoliate and clover-like in seedlings and young plants, but in the form of stiff, branched spines 1·5-2·5cm long in mature plants: slightly greyish green. The spines are soft on young shoots and readily grazed by animals.

Flowers: golden yellow, about 1·5cm long on short velvety stalks: calyx of sepals yellowish, 2-lipped and hairy. Appear mainly in spring, but often in mid-winter in mild seasons.

Fruit: pod, black, 11-20mm long, hairy, barely longer than the calyx, bursting when ripe to expel seeds.

Uses: an old use for gorse was as a fuel and, after crushing, as fodder.

BROOM
Sarothamnus scoparius

A shrub similar to gorse but without spines, the broom grows up to a height of 2m. The twigs are 5-ridged and hairless. The flowers respond to the landing of visiting insects by expelling pollen onto their undersides. The shrub is found on heaths and dunes and in woods.

Leaves: composed of three elliptic leaflets, short-stalked or stalkless and slightly hairy. Although the leaves are deciduous, the green stems give the broom an evergreen appearance.

Flowers: golden yellow, about 2cm long on stalks up to 1cm: style forming a loop. Calyx of sepals 2-lipped and hairless. Flowering May-June.

Fruit: pod, black, 2·5-4cm long; hairs on margins only.

Uses: the shoots are still used to stimulate urine production, and the long, supple branches were made into brooms for sweeping.

SPANISH BROOM
Spartium junceum

This much-branched upright shrub is a native of the dry, sun-drenched slopes of southern Europe and is much planted elsewhere for its brilliant yellow flowers. The slender, whippy twigs are green and rush-like throughout the year and the plant looks much like the previous species except that it has very few leaves or even none at all. When they are present they are very small and either oval or strap-like.
Flowers: brilliant yellow and pea-like, 2-2·5cm long and sweetly scented: in clusters near tips of branches. Much of the summer.
Fruit: a pod: green and hairy at first, becoming black and hairless: 5-8cm long with several brown seeds.

CITRUS FAMILY Rutaceae

This family includes over 1000 species, mostly of tropical and subtropical regions. The citrus trees – none of which is native to Europe – were originally from South East Asia but have been widely cultivated in warm regions (particularly the Mediterranean) since ancient times. They all have glossy evergreen leaves, whose glands secrete aromatic oils, and 5-petalled flowers.

GRAPEFRUIT
Citrus paradisi

The grapefruit is thought to be native to south-east China; it reaches a height of 10–15m.
Crown: rounded or pyramid-shaped.
Leaves: oval, with a pointed tip, smooth margins, and very broadly winged leafstalks (up to 1·5cm wide). The leaves, which often have spines in their axils, are light green when they first open, becoming darker on the upper surface.
Flowers: white, growing either singly or in clusters.
Fruit: globular, 10–15cm across, with thick smooth pale-yellow or yellow-orange rind and sweet or slightly acid-tasting flesh.

LEMON
Citrus limon

Probably native to India, this small spiny tree is the least hardy of the citrus trees; grows to a height of 6–7m.
Crown: irregular and spreading.
Shoots: reddish and bearing stout spines.
Leaves: dark green and oval, with a pointed tip and crinkly or toothed margins. The leafstalks are jointed and narrowly winged.
Flowers: fragrant, growing singly or in pairs, and developing from reddish buds. The petals are white, tinged with purple on the outside, and there are 20 to 40 stamens. In some flowers both stamens and ovaries are functional; in others only the stamens are fertile.
Fruit: egg-shaped, with a nipple-like projection at the tip; the rind is pale yellow and the flesh acid-tasting. Lemons are used principally for beverages and flavouring.

TANGERINE
Citrus deliciosa

The tangerine is a small spiny tree with distinctive fruits.
Leaves: narrow and oval, with a pointed tip and narrowly winged leafstalks.
Flowers: white, growing singly or in small clusters.
Fruit: rounded, 5–7·5cm across, and flattened or depressed at both ends. The rind – thin and bright orange – separates readily from the sweet-tasting flesh when the fruit is ripe.
It is also known as the mandarin.

CITRON
Citrus medica

This small tree was the first citrus species to be brought from the Far East for cultivation in Europe.
Crown: irregular and spreading.
Leaves: oval, with toothed margins and rounded or narrowly winged leafstalks. Short spines grow in the axils of the leaves.
Flowers: large and fragrant, developing from purple buds, with pinkish-white outer petals and many stamens.
Fruit: oblong or oval and very large (15–25cm long), with very thick rough yellow rind and pale green or yellow flesh with a sweetish or acid taste.

SWEET ORANGE
Citrus sinensis

Native to China, the sweet orange is the most adaptable of all the citrus trees to growing at lower temperatures, although it is prone to attack by pests and disease; it reaches a height of 9–13m. As well as being grown for its fruit, it is a popular ornamental pot plant.
Crown: rounded or pyramid-shaped.
Leaves: oval, 7·5–10cm long, with a pointed tip, smooth margins, and narrowly winged leafstalks. The leaves are dark green above, paler beneath.
Flowers: white and fragrant, growing singly or in small clusters.
Fruit: rounded with tough yellow, orange, or orange-red rind and sweet-tasting flesh. It is eaten fresh, made into orange juice, or used for flavouring.

SEVILLE ORANGE
Citrus aurantium

Similar to the sweet orange, this small tree can be distinguished chiefly by its fruit.
Crown: rounded or spreading.
Leaves: like those of the sweet orange, but the leafstalks are more broadly winged.
Flowers: white and fragrant, growing singly or in small clusters in the axils of the leaves.
Fruit: rounded, about 7·5cm across, but flattened slightly at both ends; the aromatic rind is orange or reddish-orange and rough, and the flesh is bitter-tasting.
Uses: the fruit is used for marmalade, beverages (including the liqueur curaçao), and confectionery (as candied peel); oil of Neroli, used in perfumery, is distilled from the flowers.

BOX FAMILY Buxaceae

BOX
Buxus sempervirens

The evergreen box, a native of southern and central Europe, parts of Britain, and North Africa, grows as a shrub or small tree on hillsides; it reaches a height of 10m. It is widely planted in parks, gardens, and churchyards, particularly as a screening and decorative hedge; it clips well and is commonly used for topiary work.
Crown: (tree form) dense and rather narrow, on a slender trunk.
Bark: thin and pale brown, patterned with small squares, becoming pale grey in old trees.
Shoots: green, covered with orange down; they are square in section.
Buds: domed, pale orange-brown, and hairy.
Leaves: hard, leathery, and oval, with a tendency to be inrolled; 1·5–3cm long. Borne on very short (1mm) stalks, they are glossy and dark green above, paler beneath.
Flowers: male and female flowers grow in loose clusters at the base of the leaves. They lack petals, consisting only of stamens (4 per flower) or styles (3 per flower).
Fruit: rounded 3-part capsules bearing the remains of the styles; they split when ripe to release small glossy black seeds.
Uses: the hard heavy close-grained wood is used for carving (e.g. chess pieces), engraving, tool handles, drawing instruments, etc.

BUCKTHORN FAMILY Rhamnaceae

BUCKTHORN
Rhamnus cathartica

Also called purging buckthorn, this rather thorny species thrives in woodland margins and hedgerows and commonly springs up on chalk and limestone hillsides when grazing is stopped. It occurs in most parts of Europe apart from the far north and the Mediterranean area. It grows as a bush or a small tree, reaching 6m in height. Many of the short twigs end in spines.
Bark: dark grey, becoming black with age.
Leaves: bright green; oval and finely toothed, often with a notch at the tip; 2-4 pairs of lateral veins. 5cm long.
Flowers: 4mm across; sweetly scented with four green petals; sexes separate. In clusters in May.
Fruit: berries 6-10mm; green and then black: poisonous, with powerful purgative action.

ALDER BUCKTHORN
Frangula alnus

Growing as a shrub or small tree and reaching 5m in height, this plant resembles the buckthorn in general appearance but differs in several details. It grows in most parts of Europe, mainly in damp woods and fens. It has no thorns.
Bark: black with clear brown pores.
Leaves: like those of buckthorn but with smooth margins and 7-9 pairs of veins.
Flowers: bisexual; 3mm across with 5 pale green petals; solitary or in small small clusters in May.
Fruit: small berries, ripening from green, through red, to black.

VINE FAMILY Vitidaceae

GRAPE VINE
Vitis vinifera

A native of S. E. Europe and S. W. Asia, the grape vine is grown throughout the warmer parts of Europe for the wine which is made from the fruit juice. It is essentially a climbing plant, with branching tendrils, but is normally cultivated as a small bush trained on wires. Many varieties are grown, most of them hybrids between American and European vines.
Leaves: palmately lobed.
Flowers: small and green, in dense clusters.
Fruit: juicy berries, green or purple according to variety.
Uses: apart from wine, the fruits of certain varieties are eaten fresh or dried (as raisins and sultanas).

QUASSIA FAMILY Simaroubaceae

TREE OF HEAVEN
Ailanthus altissima

This fast-growing Chinese tree is widely grown for shade and ornament in streets, parks, and gardens (it is naturalized in parts of southern and central Europe); it withstands pollution and is often planted for soil conservation. It reaches a height of 25m.
Crown: a tall loose irregular dome, with stout wavy up-growing branches supported on a straight trunk.
Bark: smooth, grey-brown to black, with white vertical streaks; it becomes dark grey and roughened with age.
Shoots: stout and orange-brown.
Buds: small and egg-shaped, maturing from red-brown to scarlet.
Leaves: compound, 30–60cm long, consisting of 5 to 22 pairs of leaflets. Each leaflet is narrow, oval, and pointed, 7–15cm long, with 1 to 3 large teeth on each side at the base with a large gland underneath each tooth. Borne on red stalks, 7–15cm long, the leaves are deep red when they first open, becoming deep green above and paler beneath.
Flowers: small and greenish, growing in large clusters; male and female flowers often grow on separate trees.
Fruit: twisted wings, each with a seed in the centre; 4cm long. They grow in large hanging clusters, 30 by 30cm, ripening from yellow-green to bright orange-red.

HOLLY FAMILY Aquifoliaceae

HOLLY
Ilex aquifolium

This evergreen grows as a shrub or small tree in western, central, and southern Europe; it reaches a height of 20m.
Crown: conical and spired, with upturned branches, in young trees; becomes dense and irregular with age.
Bark: smooth and silvery-grey, becoming rough and gnarled with age.
Leaves: very variable, but usually leathery and oval, 6–8cm long, with a pointed tip; glossy dark green above, bright green and matt beneath. Leaves on the lower branches have spiny margins, those higher up have wavy or smooth margins. Some cultivated ornamental varieties have variegated foliage.
Flowers: male and female flowers grow on separate trees in crowded fragrant clusters at the base of the leaves. Each flower is small (6–8mm across) and white, opening from a purple bud.
Fruit: poisonous berries, 7–10mm across, ripening from green to scarlet. Borne on a stalk 4–8mm long, each contains 3 to 4 black seeds.

MAPLE FAMILY Aceraceae

Most of the trees in this family (which contains about 150 species) are maples (genus *Acer*). Their flowers – small and greenish-yellow – grow in clusters and their fruits consist of 2 wings joined by the seeds at their bases. Most are deciduous.

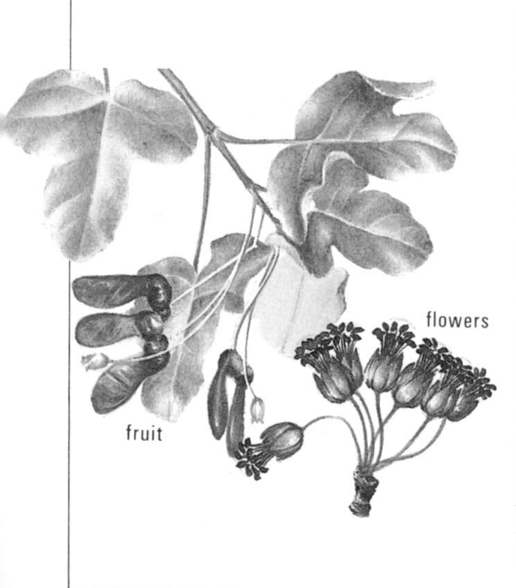

MONTPELIER MAPLE
Acer monspessulanum

This small maple, growing in dry hilly and rocky districts of southern Europe and western Asia, is sometimes planted for ornament in parks and gardens; it reaches a height of 15m.
Crown: dense and broadly domed.
Bark: dark grey to black, with vertical cracks.
Shoots: smooth, slender, and pale brown.
Buds: small (3mm), egg-shaped, and dark orange-brown.
Leaves: hard and leathery, 4 by 8cm, with 3 rounded untoothed lobes. Borne on orange-pink stalks, 4cm long, the leaves are bright green on opening, becoming dark green above and grey-blue beneath. Brilliant orange in autumn.
Flowers: grouped in flat-topped erect (later drooping) clusters.
Fruit: paired brown nutlets attached to parallel or overlapping wings—green or pinkish, each 1·2cm long and borne on a 4cm stalk.

BOX ELDER or ASH-LEAVED MAPLE
Acer negundo

This North American maple is widely cultivated in Europe for shelter and ornament, being planted in town streets, parks, and gardens; short-lived and fast-growing, it reaches a height of 20m.
Crown: irregularly domed, leaning to one side with age.
Bark: smooth and grey-brown at first, becoming darker and cracked.
Shoots: green and straight, becoming covered with purple bloom in the second year.
Buds: small, white, and silky.
Leaves: compound, up to 20 by 15cm, with 3 to 7 irregularly toothed leaflets. Borne on pale-yellow or pink stalks, 6–8cm long, the leaves are pale green, but the colour varies in ornamental varieties.
Flowers: male and female flowers grow on separate trees in hanging clusters before the leaves open.
Fruit: the wings, 2cm across, are set at an acute angle; pale brown when ripe, they remain on the tree after the leaves have fallen.

FIELD MAPLE
Acer campestre

Often growing as a small tree or shrub in hedgerows, especially on chalk or limestone soils, the hardy field maple is found throughout Europe, extending to southern Sweden, North Africa and northern Iran; reaches a height of 25m. It is grown for ornament and hedges.
Crown: domed and usually low.
Bark: pale brown with wide cracks or split into squares; becomes darker with age.
Shoots: brown, covered with fine hairs and, later, corky ridges.
Buds: brown and hairy; 3mm long.
Leaves: up to 8 by 12cm, with rounded lobes each with a shallow notch near the tip (the middle, largest, lobe has parallel sides or is wedge-shaped). Borne on slender green or pink stalks, 5–9cm long, the leaves open pinkish, becoming dark green above (paler beneath) and bright gold or reddish in autumn.
Flowers: grow in erect widely spaced clusters of about 10 and open with the leaves.
Fruit: horizontal wings, 5–6cm across, yellow-green tinged with crimson, ripening to brown.

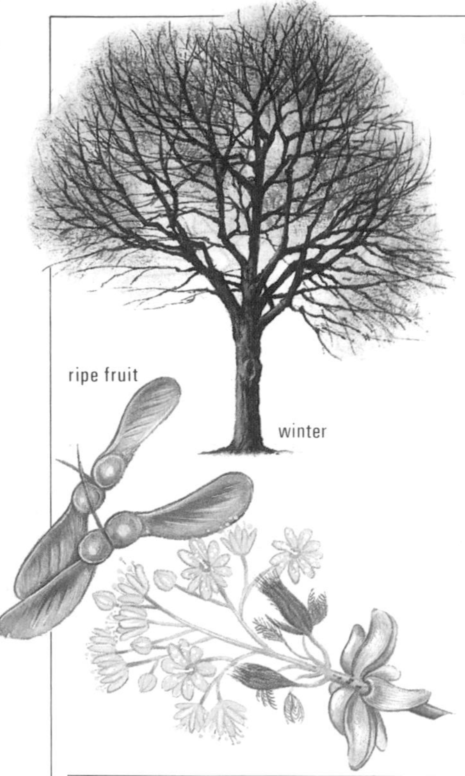

NORWAY MAPLE
Acer platanoides

The Norway maple is found all over Europe (except the extreme north); grows to a height of 30m. It is often planted as a shelter belt and for shade and ornament, particularly in city streets (since it tolerates smoke).
Crown: tall, domed or spreading, and dense, often on a very short trunk.
Bark: smooth and grey-brown, with a network of shallow ridges.
Shoots: pinkish-brown.
Buds: egg-shaped, dark red or brown (at the tips of the branches).
Leaves: 5-lobed, with the tip and teeth of each lobe ending in long slender points; 12 by 15cm. Borne on long (15cm) slender stalks containing milky sap, the leaves open rusty red, becoming bright green above and paler beneath, and turning yellow then orange-brown before falling.
Flowers: greenish-yellow and quite showy; grow in erect clusters of 30–40 that open before the leaves.
Fruit: the wings, 6–10cm across, are very widely spread; they ripen from yellow-green to brown.
Uses: the hard heavy fine-grained wood – white or greyish – is used for furniture and turned articles.

SYCAMORE
Acer pseudoplatanus

Growing wild in mountainous regions of southern and central Europe, the sycamore is widely planted for shelter, ornament, and timber (it has become naturalized in many parts); since it tolerates pollution, it is often grown in towns and cities. It reaches a height of 35m.
Crown: dense and broadly domed, with spreading branches.
Bark: smooth and grey, becoming pinkish-brown and flaking off in irregular plates.
Buds: green and egg-shaped; 8–12mm long.
Leaves: up to 18 by 26cm, divided into 5 pointed coarsely toothed lobes. Borne on reddish stalks, up to 15cm long, the leaves are orange or reddish when they first open, becoming deep green and matt above and pale blue-green beneath.
Flowers: grow in dense narrow hanging clusters of 50 to 100, 6–12cm long.
Fruit: the wings, about 3cm long and set more or less at right angles, are green tinged with red, turning brown when the fruit is ripe.
Uses: the hard yellowish-white fine-grained timber is used for furniture, turned articles (bowls, spoons, etc.), carving, textile rollers, violins and other musical instruments, and veneers.

SPINDLE TREE FAMILY Celastraceae

SPINDLE TREE
Euonymus europaeus

The attractive spindle tree is found throughout Europe (except in the extreme north), growing as a shrub or small tree in woods, thickets, and hedgerows, especially on chalk or limestone; reaches a height of 6m.
Bark: smooth and green, becoming grey or pale brown with age.
Buds: green and egg-shaped.
Leaves: oval to lance-shaped, 3–10cm long, with a pointed tip and finely toothed margins. Borne on stalks 6–12mm long, they are shiny blue-green above and paler beneath, turning yellow, russet, and crimson in autumn.
Flowers: small (1cm across), with 4 greenish-white petals and 4 stamens; they grow in loose long-stalked clusters of 3 to 8 at the base of the leaves.
Fruit: 4-lobed seed-pods, 10–15mm across, ripening from green to bright pink. When ripe they split to reveal the seeds, each covered by a fleshy orange-red coat (aril); the seeds themselves, which are poisonous, are white and surrounded by a pink seed coat.
Uses: the whitish wood is hard, smooth, and tough; it has been used for spindles, knitting needles, pegs, toothpicks, etc., and makes excellent artists' charcoal.

CASHEW FAMILY Anacardiaceae

A largely tropical family of about 600 species, but with seven species in Europe. All have small flowers, usually with 5 petals, grouped into branching clusters. Most are resinous and several are used as sources of resin and gums. They are also used for tanning leather. The European species all form thickets on the dry, stony hillsides of the Mediterranean region.

PISTACHIO
Pistacia vera

This Asiatic tree is grown in S. Europe for its edible green seeds. It reaches 10m.
Leaves: to 10cm with up to five oval greyish-green leaflets.
Fruit: egg-shaped; 2·5cm long, hard-shelled in loose bunches.

SMOKE TREE
Cotinus coggygria

Named for its smoky appearance when in fruit, this dense shrub grows to 2-5m.
Leaves: smooth and rounded, bluish-green and becoming brilliant red in autumn: 3-8cm long with clear veins.
Flowers: yellow, in loose conical clusters.
Fruit: small and rounded, brownish when ripe: in loose clusters whose stalks carry plume-like hairs, giving the tree its smoky look in late summer. Hairs vary from yellow to greyish brown.

SUMACH
Rhus coriaria

Recognized by its hairy shoots, this shrub is almost evergreen, retaining at least a few leaves for most of the winter. Reaches 3m; common by many Mediterranean roadsides.
Leaves: thick and velvety, deep green with numerous coarsely toothed oval leaflets. 20cm or more long.
Flowers: greenish, packed into dense, erect spikes.
Fruit: purplish brown and hairy; currant-sized in dense clusters.

TURPENTINE TREE
Pistacia terebinthus

Also known as the terebinth, this grey-stemmed, strongly aromatic shrub or small tree reaches 5m.
Leaves: bright green, 10cm or more long; leathery, with 3-9 oblong leaflets.
Flowers: green to purple in large clusters.
Fruit: pea-sized, bright red becoming brown.

MASTIC TREE
Pistacia lentiscus

Also known as the lentisc, this evergreen and strongly aromatic species is usually a low, spreading bush: occasionally reaches 8m.
Leaves: dark green and leathery; up to 5cm, with 6-12 narrow leaflets.
Flowers: dull red in small tight bunches: unisexual.
Fruit: pea-sized, red becoming black.
Uses: the resinous gum from stems is used as chewing gum.

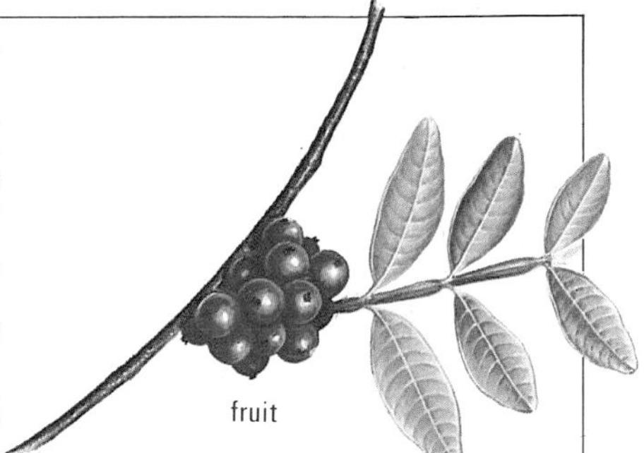

HORSE CHESTNUT FAMILY Hippocastanaceae

This small family, with about 30 species, has its headquarters in North and Central America, where the trees are commonly called buckeyes. Almost all are large and showy trees and all have the familiar digitate leaves and erect flower spikes.

HORSE CHESTNUT
Aesculus hippocastanum

Native to the Balkans, this fast-growing impressive tree is very widely planted for ornament and shade; it withstands pollution and is naturalized in some parts of Europe. It reaches a height of 30m.
Crown: tall and domed, on a short thick trunk.
Bark: dark grey-brown or reddish-brown, flaking off in scales.
Shoots: stout; grey or pink-brown.
Buds: large and pointed (2·5 by 1·5cm), shiny dark red-brown, and sticky with resin.
Leaves: compound, with 5 to 7 toothed leaflets (up to 25 by 10cm), which are pointed and broadest near the tip, arising from the same point on the stout yellow-green stalk (up to 20cm long). Bright green at first, they become darker above, yellow-green beneath, and turn gold, orange, or scarlet in autumn; they leave horseshoe-shaped scars when they fall.
Flowers: 2cm across, with fringed white petals tinged with crimson or yellow at the base; they grow in erect clusters, 15-30cm long. The flowers at the top of each cluster have only male organs (stamens) and therefore produce no fruit.
Fruit: green, globular, and spiny, splitting when ripe (brown) to reveal 1 to 3 shiny brown seeds (conkers).
Uses: the soft white wood has been used for joinery, cabinetwork, turnery, etc.: the seeds provide fodder for cattle in eastern Europe.

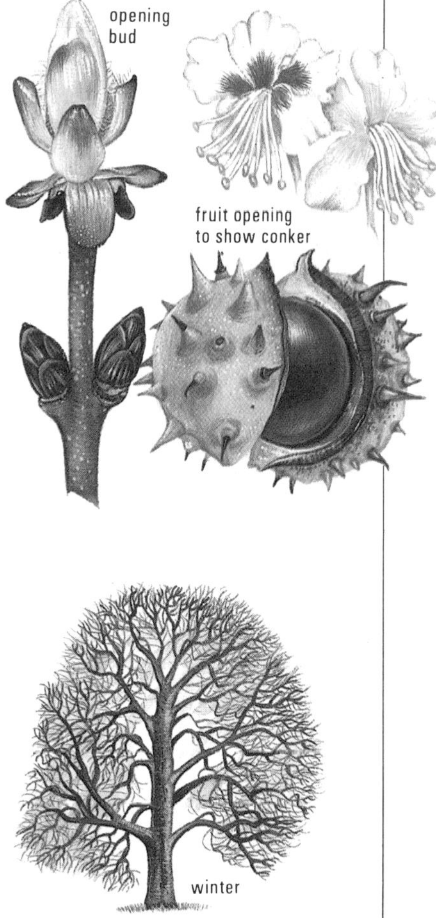

RED HORSE CHESTNUT
Aesculus × carnea

This hybrid between the horse chestnut and the American red buckeye (*A. pavia*) is widely planted as an ornamental; reaches a height of 20m.
Buds: egg-shaped, 1·5–2·5cm, but not sticky.
Leaves: similar to those of the horse chestnut but the leaflets are darker, crinkled, sometimes shiny above, and have broader, more jagged, teeth.
Flowers: red, in erect clusters 12–20cm long.
Fruit: usually not spiny; each contains 2 to 3 small dull-brown seeds.

LIME FAMILY Tiliaceae

Most of the 300 to 400 species of this family are native to tropical and warm regions. The limes, widely distributed in northern temperate regions, all have small fragrant flowers (with 5 sepals, 5 petals, and many stamens) hanging in clusters from leafy strap-like bracts; they develop into nut-like fruits, each with 1 to 3 seeds. The soft white wood is used for carving, turnery, hat blocks, piano keys, and wood pulp. The soft fibres of the inner bark are known as bast and are very strong and flexible. They were once used for making ropes and matting and even fishing nets and coarse clothing.

Lime trees are also known as lindens, and in North America they are called basswoods. They must not be confused with the limes which yield lime juice: these are citrus fruits related to the lemon and orange.

SMALL-LEAVED LIME
Tilia cordata

Growing wild throughout Europe, the small-leaved lime is also widely planted for shade and ornament, especially in avenues; it grows to 30m.
Crown: tall, dense, and domed.
Bark: smooth and grey, becoming dark grey and cracked into plates.
Shoots: red above, olive beneath.
Buds: smooth, shiny dark red, and egg-shaped.
Leaves: heart-shaped, 4–7 by 3–5cm, with finely toothed margins. Borne on yellow-green or pinkish stalks 3·5cm long, they are dark shiny green above, and paler – with tufts of reddish hairs at the bases of the veins – beneath.
Flowers: greenish-white, growing in nearly erect or spreading clusters of 4 to 15 from pale green bracts, 6cm long.
Fruit: small (6mm across), rounded, and smooth or indistinctly ribbed.

LARGE-LEAVED LIME
Tilia platyphyllos

Another widespread European lime (although not extending as far north as the small-leaved species), this tree is planted for ornament. It grows to a height of 40m.
Crown: tall and domed, with up-growing branches.
Shoots: reddish-green and hairy.
Leaves: rounded, with a pointed tip and sharply toothed margins; 6–15 by 6–15cm. Borne on hairy stalks, 2–5cm long, they are dark green and hairy above, and paler – with white hairs on the veins – beneath.
Flowers: yellowish-white, hanging in clusters of 3 to 4 from whitish-green bracts, 5–12cm long.
Fruit: rounded, 8–10mm across, each with 3 to 5 prominent ribs and densely covered with hairs.

COMMON LIME
Tilia × europaea

This tree – a hybrid between the small-leaved and large-leaved limes – is often planted for shade and ornament, especially in streets and avenues in north-west Europe; it produces numerous suckers from the base and from the trunk itself, the region from which the suckers arise often swelling up to form a prominent bulge or boss. Sucker shoots also spring in quantity from the cut ends of the branches, and large bosses develop here as well. Many limes in parks and avenues are clipped back to these bosses each year, causing them to swell even more and give the trees a strangely attractive tufted appearance.
Crown: tall and domed, with upturned branches (the lower branches are arched).
Bark: dull grey and smooth, becoming rough with a network of shallow ridges and (usually) bosses.
Shoots: green tinged with red.
Buds: reddish-brown and egg-shaped.
Leaves: heart-shaped, 6–10cm long, with toothed margins and an unequal base. Borne on green hairless stalks, 2–5cm long, they are dull green above, pale green and rather shiny beneath with tufts of buff or white hairs at the bases of the veins; the leaves turn yellow in autumn.
Flowers: yellowish-white, hanging in clusters of 4 to 10 from yellow-green bracts.
Fruit: egg-shaped, downy, and slightly ribbed; 8mm across.

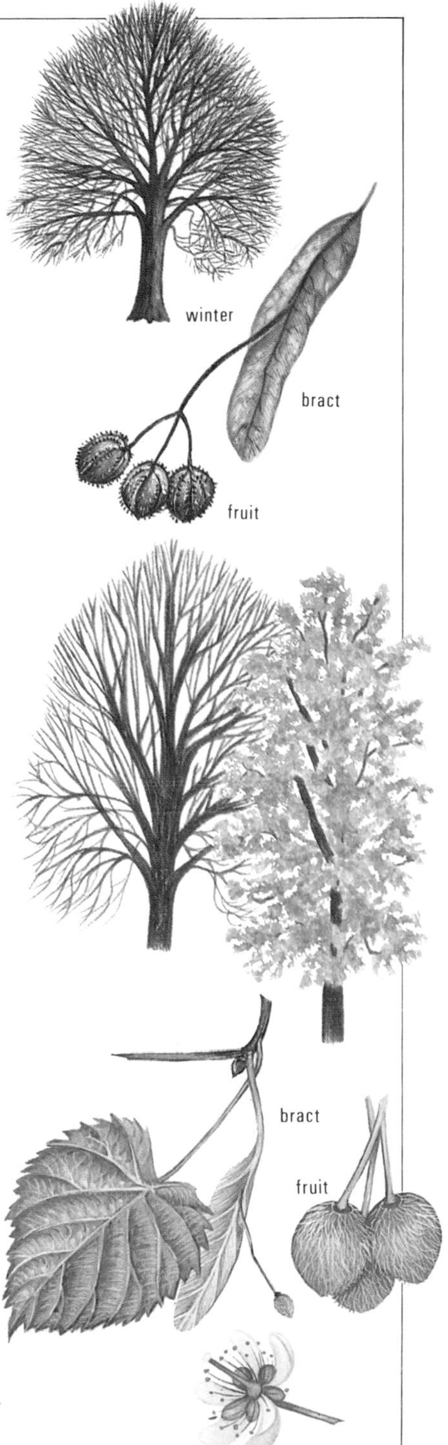

SILVER LIME
Tilia tomentosa

Native to south-east Europe (the Balkans) and south-west Asia, the silver lime is quite often planted for ornament; it grows to 30m.
Crown: broadly conical or domed, with steeply up-turned branches.
Bark: dark greenish with vertical markings, becoming grey with a network of flat ridges.
Shoots: whitish and densely downy, becoming dark grey-green above and bright green beneath.
Buds: egg-shaped and hairy, 6–8mm long; green and brown.
Leaves: rounded, with a pointed tip, toothed margins, and an unequal base; 12 by 10cm. Borne on downy 5-cm stalks, the leaves are dark green and crinkled above, pale grey and downy beneath.
Flowers: yellow-white, growing in clusters of 6 to 10 from yellow-green bracts, 9 by 2cm.
Fruit: egg-shaped and warty, 6–10mm long.

POMEGRANATE FAMILY Punicaceae

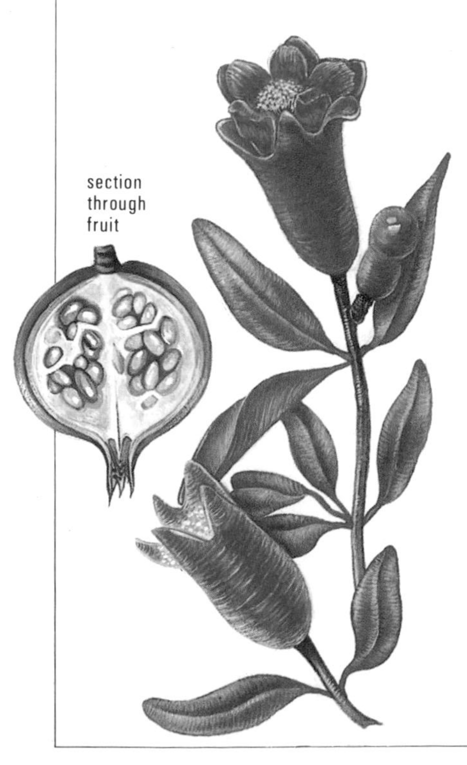

POMEGRANATE
Punica granatum

Originally from south-west Asia, this small, much-branched tree is now widely grown in southern Europe for its fruit (it is naturalized in some parts); the pomegranate is occasionally grown for its ornamental flowers in temperate regions. It reaches a height of 6m.
Shoots: angled and hairless.
Leaves: slender and shining bright green, 2–8cm long, growing on short stalks.
Flowers: orange-red and showy, 3–4cm across; the sepals are united in a tube from which 4 crumpled petals and many stamens emerge. Some ornamental varieties have double flowers.
Fruit: a large berry, 5–8cm across, with a brownish-red leathery skin enclosing a sweet or acid-tasting purple to white pulp divided into compartments containing many seeds.
Uses: the fruit is eaten raw and its juice can be drunk fresh or made into wine; the seeds are used in jams and syrups. The bark, rind, and roots of the tree were formerly used medicinally (especially as a worm powder).

MYRTLE FAMILY Myrtaceae

CIDER GUM
Eucalyptus gunnii

Native to Tasmania, the cider gum is the most commonly planted eucalyptus in north-west Europe, where it reaches a height of 20–30m.
Crown: conical at first, with upswept branches, becoming tall, domed, and heavily branched.
Bark: smooth and grey beneath pinkish-orange peeling strips.
Shoots: yellowish-white, covered with a pinkish-grey bloom.
Leaves: (mature) evergreen and oblong, 8–10 by 3–4cm, tapering to a point at the tip; borne on pale-yellow stalks 2·5cm long, these leaves are dark blue-grey above, yellow-green beneath, and smell of cider when crushed. Young leaves are rounded, 3–6cm across, and pale blue-grey.
Flowers: white and fluffy, in clusters of 3, opening from blue-white egg-shaped rimmed buds.
Fruit: white top-shaped flat-ended capsules, 5mm long

MYRTLE
Myrtus communis

A dense much-branched evergreen shrub, the myrtle grows wild in dry sunny positions, woods, and thickets in the Mediterranean region; reaches a height of 5m. It is quite widely planted for ornament.
Shoots: downy.
Leaves: oval, 2–3 by 1·5cm, with a pointed base and tip. Dark green and leathery, the leaves are dotted with glands and are very aromatic when crushed.
Flowers: sweet-scented, 2cm across, with 5 white petals and numerous stamens. Opening from globular buds enclosed in 5 shiny brown sepals, the flowers grow on long stalks from the axils of the upper leaves.
Fruit: rounded purple-black berries, 6·5mm long.
Uses: the hard mottled wood is used for turned articles and charcoal; the leaves, flowers, and fruit yield an oil used in perfumery.

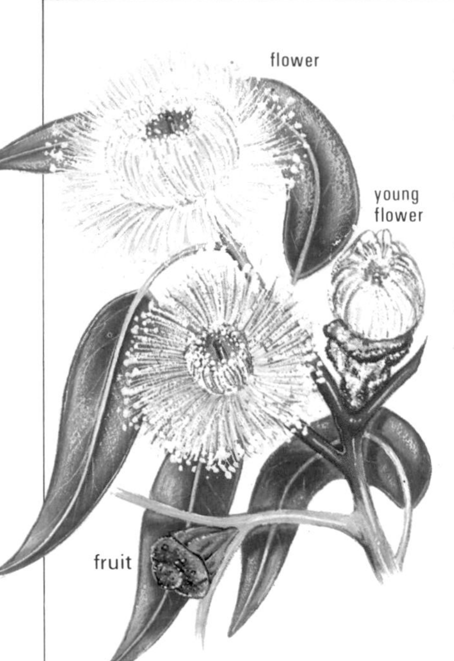

BLUE GUM
Eucalyptus globulus

This tall fast-growing evergreen from Australia is widely planted in frost-free parts of Europe for ornament and timber; reaches a height of 55m or more.
Crown: conical or domed, high, and dense, on a straight cylindrical trunk.
Bark: a pale-brown outer bark flakes off to reveal patches of grey, brown, and white smooth inner bark.
Leaves: (mature) lance- or sickle-shaped, 10–30 by 3–8cm, glossy dark blue-green and dotted with glands; (young) oblong, 10–15cm long, and pointed at base and tip; pale greyish-blue to white.
Flowers: whitish, about 4cm across, and borne singly; the petals and sepals are united to form a beaked cap and fall away to reveal numerous yellow stamens.
Fruit: large blackish top-shaped capsules, 1–1·5 by 1·5–3cm, with greyish-blue lids that open to release the seeds.
Uses: medicinal eucalyptus oil is obtained from the leaves.

ROCKROSE FAMILY Cistaceae

GUM CISTUS
Cistus ladanifer

The tallest European member of the family, this much-branched evergreen shrub may reach 3m. Its foliage smells strongly of balsam. Like the other cistuses, it revels in dry, rocky places. It forms thickets in the western Mediterranean region.
Leaves: 4-10cm long, narrow; bright and sticky above, pale and furry below.
Flowers: 5-10cm across; five white petals, each with a brown or purple patch. Each flower lasts just one day in May and June.

SAGE-LEAVED CISTUS
Cistus salvifolius

A sprawling evergreen shrub up to 1m high, flowering March-May throughout the Mediterranean region.
Leaves: oval, up to 2cm long: bright green, soft like those of sage.
Flowers: 2-4cm across; white, with prominent tuft of yellow stamens.

GREY-LEAVED CISTUS
Cistus albidus

This much-branched, upright evergreen shrub from the western Mediterranean region reaches about 1m.
Leaves: velvety grey-green; oval, 2-3cm long.
Flowers: a beautiful pink, up to 6cm across: petals often creased and lasting just one day. April-July.

TAMARISK FAMILY Tamaricaceae

TAMARISK
Tamarix anglica

A feathery shrub with slender, whip-like brown or purple branches, this plant is most frequently found on the coast or on shingly river banks. A native of S.W. Europe, it is now established in many other places.
Leaves: very small and scale-like, completely clothing the smaller twigs.
Flowers: 3mm across, pink or white; packed into dense spikes near tips of young shoots. July-September.

DOGWOOD FAMILY Cornaceae

DOGWOOD
Cornus sanguinea

Found throughout Europe (except the far north), the dogwood grows as a shrub or small tree in hedgerows, thickets, woods, and scrub. It prefers chalky soils and produces suckers freely; it reaches a height of 4m.
Bark: greenish-grey.
Shoots: dark red and very conspicuous in winter.
Buds: slender and scaleless.
Leaves: oval, 4–10cm long, with a pointed tip and prominent curved veins. The leaves turn from pale green to dark red in autumn.
Flowers: small and white, with 4 wide-spreading petals, 4 sepals, and 4 stamens. The flowers grow in flat-topped clusters, 4–5cm across, at the tips of the branches.
Fruit: globular, 6–8mm across, ripening from green to shiny black. The bitter-tasting flesh encloses a hard stone containing 2 seeds.
Uses: the wood – tough, white, and smooth – was originally used to make skewers.

SEA BUCKTHORN FAMILY Eleaginaceae

SEA BUCKTHORN
Hippophae rhamnoides

Unrelated to the true buckthorn (page 201), this densely branched, spiny shrub grows mainly by the sea, often being planted to stabilize dunes; also by mountain streams. Up to 10m.
Shoot: spine-tipped and clothed with golden brown scales.
Leaves: slender, 1-6cm long; greyish green at first, owing to coat of silvery hairs, becoming brighter and hairless.
Flowers: very small and green but clothed with brown scales; densely clustered; unisexual.
Fruit: clustered orange berries 6-8mm.

HEATHER FAMILY Ericaceae

This is a family of 2,000 or so shrubs, together with a few small trees, spread through most of the cooler parts of the world, including the tropical mountains. Most, including the heather itself, are low-growing. Just a few of the taller species are described here.

STRAWBERRY TREE
Arbutus unedo

The small evergreen strawberry tree grows naturally in thickets, woods, and dry rocky places of south and south-west Europe as far north as south-west Ireland; it is sometimes planted for shelter and ornament. It reaches a height of 12m.
Crown: low, dense, and rounded, with up-growing branches supported on a very short trunk.
Bark: dark reddish, later peeling off and forming grey-brown ridges.
Shoots: hairy; pinkish above, pale green beneath.
Leaves: lance-shaped, with sharply toothed margins; 5–10 by 2–3cm. Borne on hairy pinkish stalks, 5–7mm long, the leaves are shiny dark green above, paler green beneath.
Flowers: white (tinged with green or pink) and flask-shaped, 8 by 8mm, growing in hanging clusters 5cm long. The flowers appear in autumn, at the same time as the ripe fruit from the previous year.
Fruit: rounded, 2cm across, with a rough warty skin, ripening from yellow to scarlet. Though edible, the fruit does not have a pleasant flavour.

RHODODENDRON
Rhododendron ponticum

A native of southern Europe, this huge spreading evergreen shrub is widely planted for its beautiful flowers and has become naturalized in many parts of the continent, especially on sandy soils. It reaches 4m or more and forms impenetrable thickets in many woods.
Bark: Brownish grey, usually with numerous shallow fissures breaking it into small rectangular blocks.

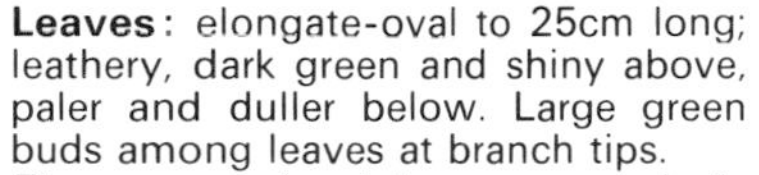

Leaves: elongate-oval to 25cm long; leathery, dark green and shiny above, paler and duller below. Large green buds among leaves at branch tips.
Flowers: pale pink to mauve, bell-shaped up to 7cm across: 10 long curved stamens and a single style project from the bell. In dense rounded clusters in June.
Fruit: a dry, brown capsule.
Several other species are cultivated in parks and gardens, but they do not usually become naturalized. One exception is the yellow rhododendron (*R. luteum*), a deciduous species from S. E. Europe. It bears deep yellow flowers on leafless shoots in spring, before the rather pale green and sticky leaves appear. The flowers have only five stamens. The plant seeds readily and also spreads by means of suckers from the roots.

LUSITANIAN HEATH
Erica lusitanica

A native of the western Mediterranean region and the Iberian Peninsula, this densely branched evergreen shrub reaches 3-4m. It grows thickly in open woods and on exposed hillsides, and when in flower stands out like a hazy white cloud against the other shrubs.
Leaves: light green and needle-like, 3-5mm long in whorls on the hairy stems.
Flowers: white with a delicate tinge of pink (especially noticeable when in bud) and a red stigma; almost cylindrical, 4-5mm long, in dense elongated clusters towards the ends of the branches. Mid-winter to mid-summer. Fragrant.
The very similar tree heath (*E. arborea*) grows throughout the Mediterranean area. It is usually taller and has darker leaves, but it is best distinguished by its white bell-shaped flowers and yellow stigmas.

HONEYSUCKLE FAMILY Caprifoliaceae

WAYFARING TREE
Viburnum lantana

The wayfaring tree is found over most of Europe (except the extreme north), growing at the edges of woods, in thickets, and in hedges, especially on chalk or limestone soils; reaches a height of 6m.
Shoots: covered with greyish down.
Buds: lack scales.
Leaves: heart-shaped and wrinkled, 5–12cm long, with toothed margins and densely covered with down on the lower surface: thev are borne on short downy stalks.
Flowers: all are white, small (6mm across), and fertile, grouped in flat-topped clusters, 6–10cm across.
Fruit: small, oval, and flattened, in flat-topped or domed clusters, ripening unevenly from green to red and finally black.

GUELDER ROSE
Viburnum opulus

Native in most parts of Europe, the guelder rose grows in woodlands, scrub, hedgerows and thickets; especially on damp soils. It reaches a height of 4-5m. Cultivated forms, planted for their flowers, are sterile.
Crown: spreading.
Shoots: smooth and angled.
Buds: greenish-yellow and scaly.
Leaves: 3- to 5-lobed, 5–8cm long, with deeply toothed margins and 2 glands at the base. Borne on greenish-red stalks, 3–4cm long, each with pointed stipules at its base, the leaves are downy at first, becoming smooth above and turning scarlet in autumn.
Flowers: white, in dense flat-topped clusters 5–10cm across. The outer flowers are large (2cm across) and sterile (without stamens or ovaries) and serve to attract insects to the flower head; the inner flowers are smaller (6-8mm across) and fertile.
Fruit: red, translucent, and berry-like, 8mm across.

ELDER
Sambucus nigra

The elder grows throughout Europe as a shrub or small tree of woods, hedges, scrub, and waste ground; it reaches a height of 10m.
Crown: irregular and much-branched.
Bark: greyish-brown, with thick corky ridges and deep cracks.
Shoots: stout and grey, with corky pores and thick white pith.
Buds: scaly; brownish-red or purple.
Leaves: compound, with 2 to 3 pairs of leaflets and one terminal leaflet. Each leaflet is oval to lance-shaped, 3–9cm long, with sharply toothed margins and a pointed tip. Aromatic.
Flowers: small (5mm across) and creamy white, grouped into flat-topped clusters 10–20cm across.
Fruit: berry-like and juicy, in large clusters, ripening from green to black.
Uses: the fruit, rich in vitamin C, is made into wine, jam, and jelly; the flowers can also be brewed to make a beverage. The hard, whitish wood is used for small articles.

BUDDLEIA FAMILY Buddlejaceae

COMMON BUDDLEIA
Buddleja davidii

Known as the butterfly bush because of its great attraction for butterflies, the buddleia is a native of China, but widely grown in gardens and naturalized in many waste places, including railway yards and gravel pits and many urban areas. It was abundant on bombed sites after the war. A quick-growing and rather straggly shrub, it reaches 5m or more: new shoots can grow 2m in a season.
Leaves: lanceolate, to 25cm; dark green above, pale and woolly below; lightly toothed.
Flowers: tubular, pale mauve to deep purple with an orange throat; about 1cm long with four lobes and abundant nectar. In long, tapering spikes to 30cm in summer.
Fruit: a slender capsule with many tiny seeds.

OLIVE FAMILY Oleaceae

This widely distributed family of trees and shrubs (400 to 500 species) includes many ornamental trees (such as the lilac and jasmine) as well as the economically important olive and ash. Their flowers usually contain only 2 stamens.

OLIVE
Olea europaea

A slow-growing long-lived evergreen, the olive has been widely cultivated for fruit since ancient times in its native Mediterranean region; reaches a height of 15m. It grows wild in dry rocky places and is also planted in gardens for shade.

Crown: spreading (bushy in wild forms), supported on a twisted gnarled trunk.

Bark: smooth and grey at first, becoming darker and deeply pitted with age.

Shoots: covered with silvery scales.

Leaves: leathery and lance-shaped, 2–8cm long, dark green above and silvery-grey beneath. Wild trees often have small oval leaves.

Flowers: small and white, each with 2 yellow-tipped stamens, growing in dense clusters in the axils of the leaves. Some flowers have no pistils.

Fruit: egg-shaped, 1–3·5cm long, containing a single large seed; olives ripen from green to black or brownish-green.

Uses: the fruit is used as food and yields a high-quality oil (olive oil) used in cooking, medicine (as a lubricant, in ointments, and as a mild laxative), and in soap-making; the residue of the fruit (after the oil has been extracted) is used as cattle food. The wood, which is very hard, is used for carving, cabinetwork, fuel, and charcoal.

flowers

unripe fruit

ripe fruit

COMMON ASH
Fraxinus excelsior

Widely distributed in Europe and south-west Asia, the common ash is found in woods and scrub (it grows particularly well on moist alkaline soils); reaches a height of 40m. It is valued for its timber and also grown for ornament in parks, churchyards, etc.

Crown: tall and rounded, with steeply up-growing branches.

Bark: pale grey; smooth at first, it later develops a network of ridges.

Shoots: stout; greenish-grey with white markings.

Buds: conspicuously black, squat, and angled.

Leaves: compound, 20–35cm long, with 4 to 6 pairs of leaflets and one terminal leaflet. The leaflets, up to 12cm long, are stalkless (or nearly so) and pointed, with sharply toothed margins; dull green above and pale and downy beneath, they turn yellow in autumn.

Flowers: usually, male and female flowers grow on separate trees, but some trees bear both male and female flowers and others have hermaphrodite flowers. The flowers of both sexes are small and purplish, growing in small dense clusters that appear well before the leaves.

Fruit: strap-shaped keys, 2·5–5cm long, each with a notched tip. They ripen from shiny deep green to brown and remain on the tree after the leaves have fallen.

Uses: the pale tough elastic timber is used for sports equipment (oars, hockey sticks, etc.), tool handles, furniture, walking sticks, pegs, etc.; it also makes good fuel and charcoal.

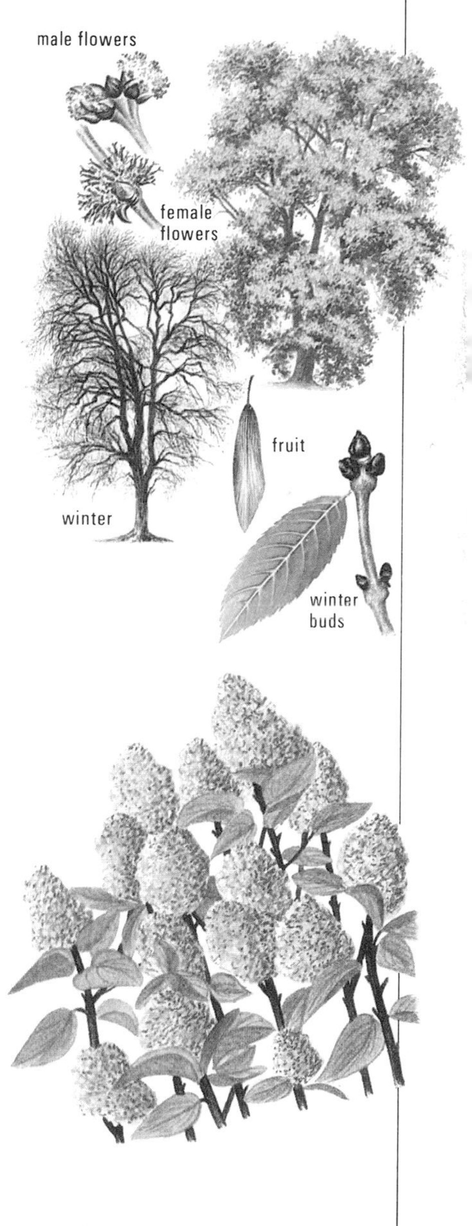

LILAC
Syringa vulgaris

Famed for its attractive and sweetly scented flowers, the lilac is a deciduous shrub or small tree, normally 3-4m high but occasionally reaching 7m. It is a native of rocky hillsides in the Balkans, but has been introduced to many places and has often become naturalized. Cultivated varieties, with flowers ranging from white to deep purple, are grown almost everywhere.

Leaves: pale green, smooth, and hairless; heart-shaped and untoothed: 4-10cm long.

Flowers: tubular or funnel-shaped, about 5mm across: normally pale mauve (lilac) in the wild but occasionally white. Very fragrant. In dense pyramidal heads up to 20cm long.

Fruit: a pointed leathery capsule with two compartments and two winged seeds in each.

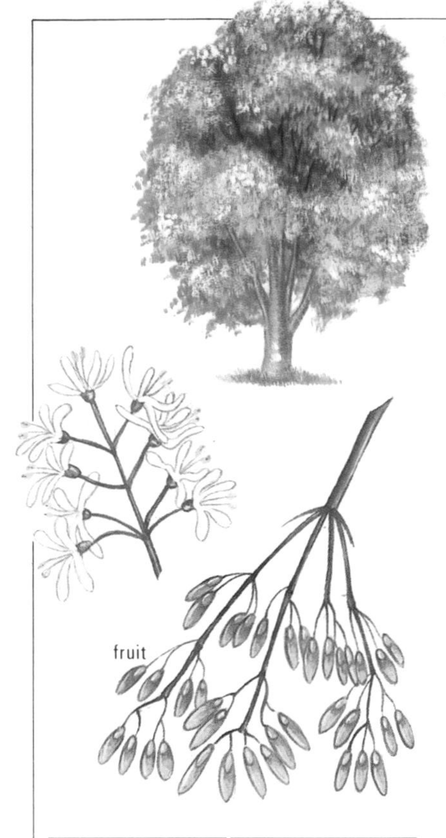

MANNA or FLOWERING ASH
Fraxinus ornus

The manna ash is native to south, south-east, and central Europe, where it grows in dry rocky regions, woods, and thickets; reaches a height of 20m. It is widely planted as an ornamental in parks, and is cultivated for commercial purposes in Italy.
Crown: rounded, with curving branches.
Bark: grey and very smooth.
Shoots: olive-green.
Buds: domed and densely covered with grey down; enclosed in 2 dark outer scales.
Leaves: compound, 25–30cm long, with 2 to 4 pairs of leaflets and one terminal leaflet. The leaflets (10 by 3·5cm) are stalked, lance-shaped, irregularly toothed, and covered with brownish or white down on the lower surface.
Flowers: white and fragrant, with very narrow 6-mm long petals, growing in dense conical clusters, 15 by 20cm, at the tips of the branches.
Fruit: slender wings ('keys'), 1·5–2·5cm long, each with a seed at its base. The keys hang in bunches and ripen from green to brown.
Uses: the branches yield a sap (manna), which hardens to form a gum used in pharmacy.

PRIVET
Ligustrum vulgare

The privet is a rather loosely branched, spreading shrub growing wild in most parts of Europe, especially on chalk and limestone soils. It is particularly common on wooded slopes and woodland margins. Reaching 4m, it is essentially deciduous, but some of the leaves, especially at the top, are reluctant to fall and may remain on the plant all winter.
Bark: smooth and grey, often reddish on young stems.
Leaves: elongate-oval, 3-6cm long; leathery with smooth margins; deep green, often becoming bronze in autumn.
Flowers: creamy white, 3-4mm across, funnel-shaped with four spreading lobes: in dense terminal spikes May-June. Strongly scented.
Fruit: a shiny black berry 6-8mm in diameter, with oily flesh. Poisonous.
The garden privet used for hedging is *L. ovalifolium* from China and Japan. It has broader leaves and tends to be more evergreen. A yellow-leaved variety is often planted.

CHAPTER THREE

FERNS AND MOSSES

The ferns and mosses are flowerless plants which reproduce by scattering minute spores instead of seeds. The ferns and the related horsetails and clubmosses form the group known as the Pteridophytes, while the mosses and liverworts are known as Bryophytes. Both are very ancient groups and were thriving well over 300 million years ago. Giant horsetails and clubmosses dominated the coal forests at that time although today's species are all rather small and insignificant plants.

Ferns

Ferns normally have underground stems so all that we usually see of them are their leaves or fronds. The European species are almost all perennials. New fronds spring from the top of the stem each year; they are tightly coiled at first and then uncoil in a very characteristic way. The fronds are often very large and, as in the Bracken (see page 229), divided into numerous leaflets. The plants' spores develop in little capsules born on the fronds – often on ordinary fronds, but sometimes on special fertile fronds or on special parts of the fronds. The capsules are carried in clusters known as sori which appear as rusty patches when the spores are ripe. In their early stages, the sori are protected by small flaps of tissue (see page 227); the shape of these flaps, together with the general arrangement of the sori on the fronds, is of great value in classifying and identifying the ferns. The overall shape of the fronds and the degree of division are also very useful in this respect.

Clouds of spores are scattered in dry weather, but they do not grow directly into new fern plants. Given suitably moist conditions, each spore grows into a thin green plate called a prothallus. Roughly heart-shaped and a few millimetres in diameter, the prothallus becomes anchored to the ground by a few hair-like roots. It is easy to grow prothalli by shaking some spores on to a dish of sand or peat and keeping it moist for a few weeks.

Sex organs are borne on the prothallus and, in damp conditions, male cells are released to swim to the female cells. The latter remain embedded in the prothallus. Each fertilized female cell eventually grows into a new fern plant, drawing food from the prothallus at first and then putting down its own roots. Because moisture is essential for the reproductive process, ferns are abundant only in damp habitats, although individual plants

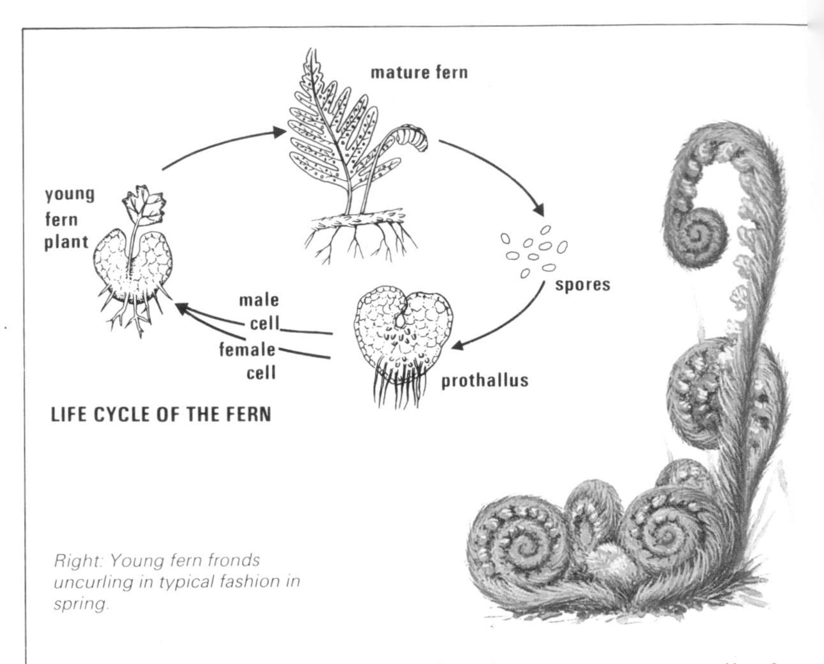

Right: Young fern fronds uncurling in typical fashion in spring.

may be able to survive in quite dry places – on stone walls, for example – once they get established. The vast majority of the 10,000 or so known ferns grow in the tropics. Only about 100 occur in Europe.

Horsetails and Clubmosses

The horsetails and clubmosses are built quite differently from the ferns (see pages 230-1). Their leaves are very small or absent and they often carry their spores in simple cones. Their life cycles are very similar, however, with a spore-producing generation alternating with a prothallus carrying sex organs. The three groups clearly belong together in the Pteridophyta.

The horsetails, of which there are only about 25 living species, scatter their spores from cones; their prothalli are rounded green cushions, up to 3 cm in diameter and bearing numerous upright lobes. Most of the 1,000 or so species of clubmosses live in the tropics; only about ten are found in Europe. They carry their spores in cones or in capsules attached to normal leaves. The prothalli of most European species are subterranean and contain no chlorophyll. They cannot make food and depend entirely on a close relationship with soil-living fungi. They grow very slowly and it may be several years before they are ready to produce their sex organs.

A horsetail prothallus with new horsetail plants springing from the cushion. The mossy 'fronds' of the prothallus are where much of its food is made.

Mosses and Liverworts

These are all low-growing plants with no true roots. Mosses all have slender upright or trailing stems with small leaves and they commonly form dense cushions or carpets on the ground. Their stems never contain water-carrying tubes, such as are found in the ferns and their relatives and in the seed-bearing plants. Most liverworts have stems and leaves like those of the mosses, but some of them look more like seaweeds, with just a flat green thallus undivided into stem and leaf (see page 234). The moss or liverwort plant is really the equivalent of the fern prothallus, for this is where the sex organs are produced. Fertilization occurs in damp conditions, as with the ferns, and the fertilized female cell grows into a stalked spore capsule. Although attached to the original plant, the capsule is really a separate organism and equivalent to the whole fern plant. The liverwort spore capsule is very much simpler than that of the moss (see page 232), but apart from this difference it is not always easy to separate the two groups. Identification of the species involves close study of the leaf shapes and their arrangement on the stem and cannot be done without a good hand lens. When studying mosses, the shape of the spore capsule is also a useful identification guide.

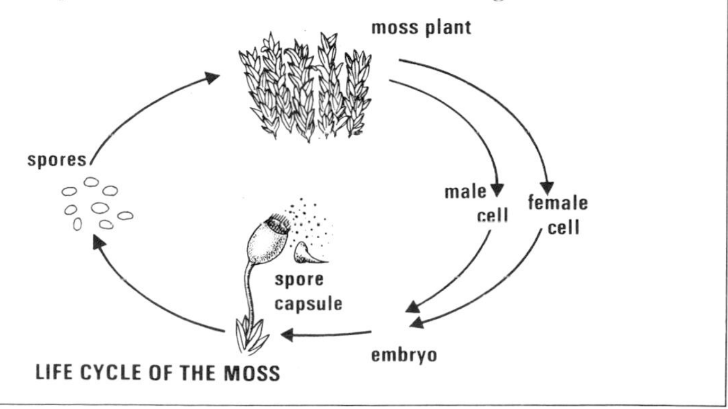

LIFE CYCLE OF THE MOSS

Ferns

European ferns are all herbaceous plants with leaves springing from underground stems. Most species have a short and stocky stem with a crown of fronds at the top, but some have creeping rhizomes which spread through the soil and send up fronds at intervals. The leaves of some species die down each autumn, but many survive until the next year's leaves are well developed and sometimes even longer.

The ferns are essentially plants of damp habitats and are much more abundant in western Britain than in the east. This is especially true of those species that grow on rocks and walls and as epiphytes on trees, for they are very dependent on a moist atmosphere.

The shape of the fronds and the position of the sori are important in the identification of the species. The following species belong to several different families.

ROYAL FERN
Osmunda regalis

A large and striking fern of fens, damp woodlands and other wet places. Widely distributed in Europe and elsewhere, but not common.
Frond: up to 3 m tall and 1 m wide, with numerous oblong leaflets (pinnules) up to 6 cm long. Outer fronds all sterile; inner ones with normal pinnules on lower part and small brown fertile pinnules higher up.
Sori: densely clustered around fertile pinnules, which have no green blade. Spores ripe June-August.

ADDER'S TONGUE
Ophioglossum vulgatum

A small and very unfern-like fern, easily overlooked in damp grassland and woodland. All Europe, but less common in the north.
Frond: each plant produces just one stalked, oval frond each year, normally 5-20 cm long and undivided; easily mistaken for a plantain leaf, but lacks the prominent veins. In older plants a slender spike arises from the base of the blade; clusters of large spore capsules develop on this spike. Spores ripe May-August.

MALE FERN
Dryopteris filix-mas

One of the commonest woodland ferns, growing throughout Europe; often in hedgerows. The fronds grow in a ring, leaning out from the base and looking like a giant shuttlecock. One of several similar species.

Frond: up to 1·5 m long; stalk clothed with pale brown scales. Divided into 20-35 tapering lobes (pinnae); longest pinnae near the centre. Each pinna is again divided into a number of smaller lobes (pinnules), although these are not completely separated in the narrower part of the pinna. Pinnules are toothed and bluntly-rounded at the tips.

Sori: in 2 rows under each well-formed pinnule; up to 6 in a row. Each sorus covered by a heart-shaped flap which shrivels as spores ripen July-August.

HARD SHIELD FERN
Polystichum aculeatum

A fern of the woodland and hedgerow, found in most parts of Europe. Resembles Male Fern in growth form, but tougher and with several detailed differences.

Frond: up to 1 m long; stalk clothed with brown scales. As many as 50 pinnae on each side, each pinna with up to 15 pinnules on each side. Pinnules are toothed and bristly and the lowest one on each pinna is much larger than the others.

Sori: in 2 rows under each well-formed pinnule, but covered with a round flap (not heart-shaped as in the Male Fern). Spores ripe July-August.

WALL RUE
Asplenium ruta-muraria

Grows in the same kinds of places as the next species. Both prefer limestone rocks and are therefore quite happy with their roots in the mortar of old walls.

Frond: 3-15 cm long; stalk longer than blade. Latter is roughly triangular and divided into leaflets which may be round, triangular, or diamond-shaped.

Sori: linear, about 2 mm long, on underside of basal part of leaflet: merging into a single mass when mature. Spores ripe June-October.

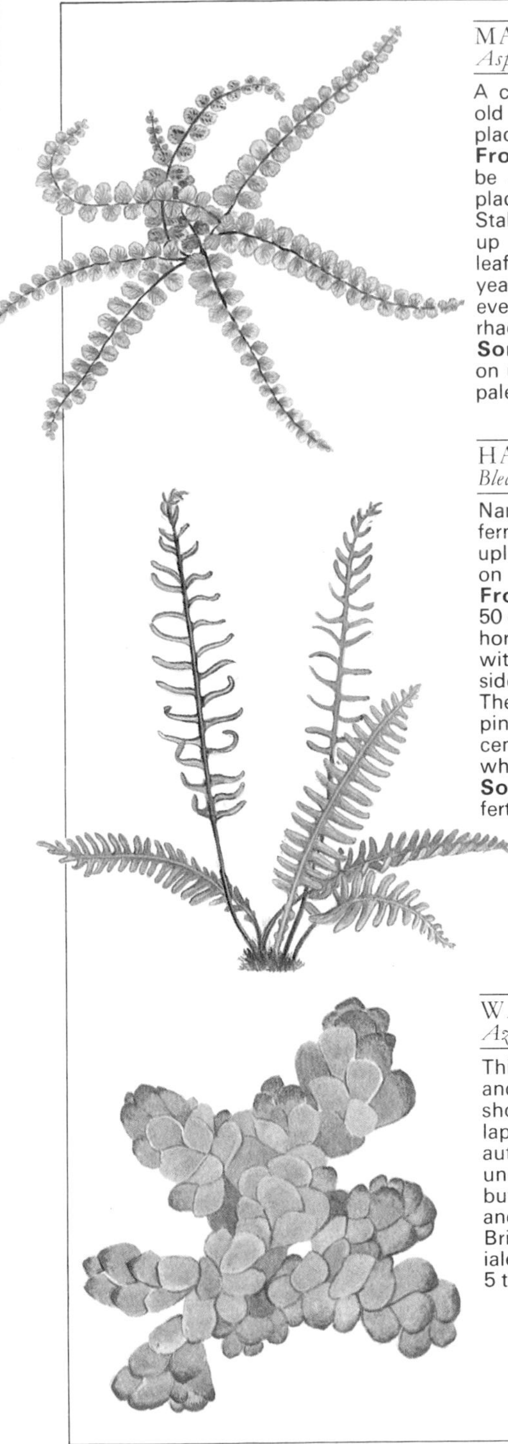

MAIDENHAIR SPLEENWORT
Asplenium trichomanes

A common fern of rock crevices and old walls in damp climates and shady places. Almost world-wide.
Frond: usually up to 20 cm long (may be as much as 40 cm in really moist places); hugs wall or rock quite closely. Stalk and mid-rib (rhachis) black, with up to 40 pairs of deep-green, oval leaflets. The latter fall in their second year, starting from the bottom and eventually leaving just the bare, wiry rhachis which is shed later.
Sori: oblong or linear, 1-2 mm long, on underside of leaflet and covered by pale flaps. Spores ripe May-October.

HARD FERN
Blechnum spicant

Named for its tough fronds, this is a fern of woods and heaths, especially in upland regions; most of Europe, but not on limestone.
Frond: 2 kinds. Sterile fronds, up to 50 cm long, spread more or less horizontally and are vaguely comb-like with up to 30 linear pinnae on each side. They persist for 2 years or more. The fertile fronds have more slender pinnae and stand ladderlike in the centre of the plant. They die away when they have shed their spores.
Sori: linear, under the pinnae of the fertile fronds. Spores ripe June-August.

WATER FERN
Azolla filiculoides

This little plant floats on ponds, canals and slow-moving backwaters. The short stems bear lots of small, overlapping leaves which often turn red in autumn. Spores are borne on the undersides. A native of North America, but now widely distributed in southern and central Europe, including southern Britain. It belongs to the order Salviniales. (The illustration is approximately 5 times life size.)

BRACKEN
Pteridium aquilinum

The commonest of all ferns, found all over the world. Basically a woodland plant, it also grows on heathland and on many hillsides, but not in the most exposed areas because young fronds are damaged by frost and cold winds. Absent from the wettest soils and from limestones. It spreads rapidly by means of an extensive rhizome system and is a troublesome weed in many hill pastures. Because it spreads as much or more by the rhizomes as by spores, it is less confined to moist places than most other ferns.
Frond: grows singly from the rhizome, to a height of 4 m (usually much less) Blade is triangular and finely divided on a long stalk; early stage shaped like a shepherd's crook and clothed with soft brown hair.
Sori: in continuous lines around the edges of the leaf segments, protected by in-curved margin. Spores ripe July-August.

HART'S TONGUE FERN
Phyllitis scolopendrium

An easily recognized fern of woodland and hedgerow, and also of rocks and walls in damp and shady places. Most of Europe, but rare in the north.
Fronds: up to 60 cm long, in dense tufts. Stalk purplish-black with brown scales at first. Blade tongue-shaped and undivided; very pale green at first, becoming bright and very glossy and then gradually darker. Fronds persist until well into their second year.
Sori: linear, forming long strips along the veins on the underside of the fronds; covered by a flap on each side at first, but pushing these aside as the capsules mature. Spores ripe July-October.

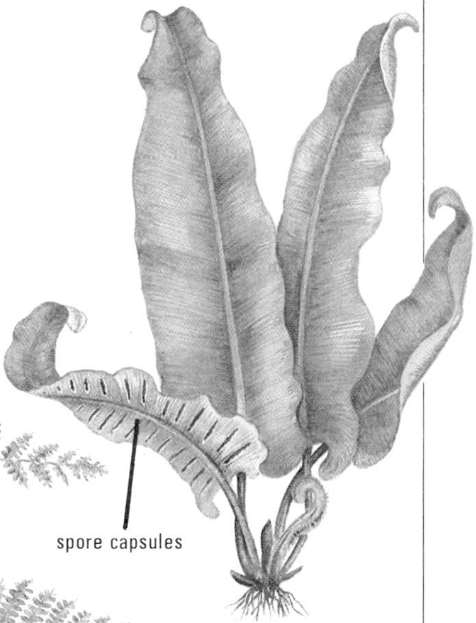

spore capsules

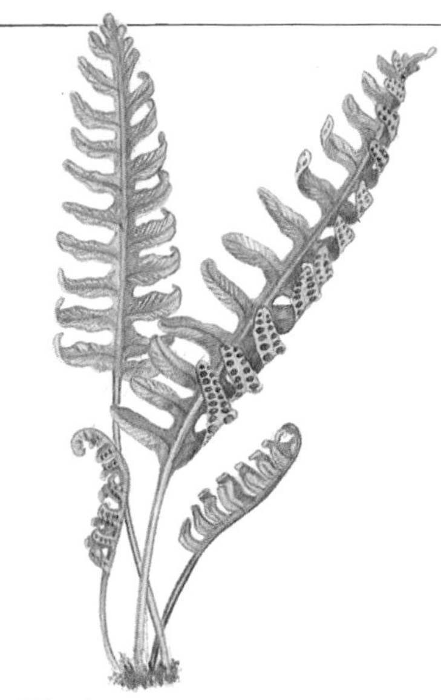

COMMON POLYPODY
Polypodium vulgare

One of 3 very similar species, this is a woodland fern, often growing epiphytically on tree trunks and branches; also on rocks and walls in wetter areas. It has a scaly, creeping rhizome, often on the surface, from which fronds arise singly.
Frond: up to 45 cm long, with up to 25 lobes (pinnae) on each side; pinnae often not completely separated.
Sori: rounded, 1-3 mm across, in 2 rows under pinnae. Spores ripe June-September.

Clubmosses

Although they resemble mosses in their slender stems and small leaves, these plants (Lycopsida) are more closely related to the ferns. Spores are borne in capsules at the bases of certain leaves, which are often grouped into club-shaped cones. Most of the 1000 or so species are tropical.

FIR CLUBMOSS
Lycopodium selago

A plant of upland moors and rocky slopes. Its tufted upright stems reach 25 cm and are clothed with stiff, sharply-pointed leaves. Pale spore capsules grow in the axils of many of the leaves without cone formation. Spores ripe June-August.

STAG'S-HORN CLUBMOSS
Lycopodium clavatum

A widely distributed plant of upland moors, grasslands, and open woods; prefers acid soils. Its much-branched, wiry stems creep over the ground, rooting at intervals and reaching lengths of 1 m or more. Fertile branches bear 2 slender yellow cones. Spores ripe June-September.

Horsetails

Once major components of the great coal forests, these plants (Sphenopsida) are now reduced to about 25 small species. All are perennials with underground rhizomes and erect stems, often with whorls of slender branches. Leaves usually reduced to small scales on the stems. Spores are borne in cones – either at the tips of the ordinary stems or on special brown shoots.

GREAT HORSETAIL
Equisetum telmateia

Up to 2 m high in damp, shady places in most parts of Europe. Sterile stems white with collars of black-tipped scale leaves and dense whorls of bright green branches; die down in autumn. Fertile shoots pale brown, up to 40 cm high in spring; cone 4-8 cm long. Spores scattered April, after which fertile shoot dies. Field Horsetail (*E. arvense*) is smaller and with green sterile stems.

WATER HORSETAIL
Equisetum fluviatile

Up to 1·5 mm in swampy ground and at the edges of lakes and ponds all over Europe. Stem green, sometimes unbranched, but often with a few short branches in upper region; collars of black-tipped scale leaves. Cones, 1-2 cm long, at tips of normal stems; spores ripe June-July.

fertile shoot

Water Horsetail

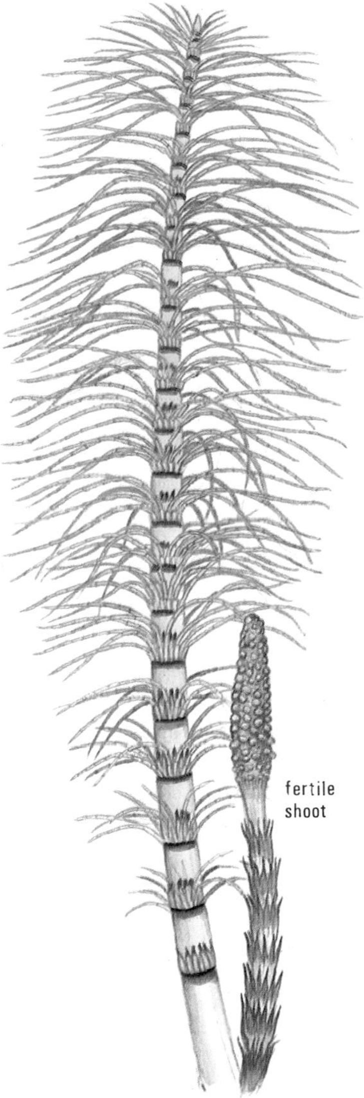

Great Horsetail

Mosses and Liverworts

Mosses (Class Musci) all have upright or creeping stems clothed with small leaves. Most liverworts (Class Hepaticae) have a similar build, but some (see page 234) are flat and seaweed-like. Leafy liverworts usually grow in turf or among mosses and are easily overlooked, but may be distinguished from mosses by their lobed leaves without mid-ribs: moss leaves are unlobed and usually have mid-ribs. The main difference between the groups lies in the spore capsule. That of a moss is urn-shaped, with a detachable lid and a system of teeth which allow the spores to escape only in dry weather. Liverwort capsules are simple spheres that split open at maturity and look like little four-rayed stars. Some mosses and most liverworts also reproduce by scattering tiny flakes of tissue, called gemmae, which grow directly into new plants. Gemmae may be borne in little cups and splashed out by rain drops.

COMMON HAIR MOSS
Polytrichum commune

One of several similar species, forming extensive dark green mats on moors and bogs. Stems upright, reaching 20 cm or more. Leaves spear-like, 8-12 mm long with toothed edges. Separate male and female stems, the tips of male shoots being flower-like when male organs are ripe.
Spore capsule: 4-sided and box-like; horizontal when ripe; covered with a furry golden hood in early stages.

SILKY WALL FEATHER MOSS
Camptothecium sericeum

Forms extensive mats on walls and tree stumps; easily recognized by the shiny yellow tips of main shoots. Secondary shoots greener, forming a deep 'pile' in centre of mat. Leaf is a narrow triangle. Separate male and female plants.
Spore capsule: an erect yellowish-green cylinder 2-3 mm long.

BRYUM CAPILLARE

A very common moss forming neat cushions on tree stumps and rocks and especially on old walls. Stems up to 5 cm high. Leaves oblong, ending in a hair-like tip and strongly twisted when dry. Separate male and female plants.
Spore capsule: pear-shaped and drooping; bright green, becoming brown.

THUIDIUM TAMARISCINUM

This beautiful woodland moss is easily recognized by its regular, Christmas-tree-like branching. Leaf is heart-shaped or triangular, densely clothing the upper parts of the black stems. Separate male and female plants. It prefers shady places on heavy soils.
Spore capsule: sausage-shaped, borne horizontally on a red stalk, but rarely developed.

EURHYNCHIUM PRAELONGUM

Abundant in shady places on heavy soils; also clothes stones and piles of rubble. It is a trailing moss with fairly thick, but weak stems. Leaves on main stems are heart-shaped; those on side branches much narrower. Separate male and female plants.
Spore capsule: egg-shaped, 1-3 mm long with a long beak when young; reddish-brown when ripe. Uncommon.

SPHAGNUM PAPILLOSUM

This is one of the commoner bog mosses that form dense rounded hummocks all over the surface of the blanket bogs of the north and west. It also occurs in many lowland bogs and it is one of the most important peat-forming plants. The hummocks consist of long stems, each bearing whorls of robust, blunt-tipped branches that form conspicuous rosettes at the surface. Small triangular leaves clothe the branches and, as in all bog mosses, give the plant its spongy nature and allow it to carry its own water supply. As the lower parts of the stems die they are gradually converted into peat. Living parts at the surface are normally yellowish brown.
Spore capsule: male and female plants are separate and spore-capsules are rarely formed, but when they do appear they are like tiny matches sticking up from the rosettes.

MARCHANTIA POLYMORPHA

A common thalloid liverwort of wet paths, river banks, greenhouses and other damp places: easily identified by the goblet-shaped gemma cups and the hexagonal pattern on the upper surface. Separate male and female plants, bearing sex organs on raised parasol-like structures. Simple spore capsules develop on underside of female parasols.

PELLIA EPIPHYLLA

A large thalloid liverwort carpeting sizeable areas of stream banks and other moist, shady places, especially on acid soils. Its irregularly-branched, shiny thallus is up to 1 cm wide and may be several centimetres long. No gemma cups. Male and female organs on the same plant, female organs under small flaps. Greenish-black spore capsules emerge from the flaps in early spring and rise on pale stalks to 5 cm.

CRESCENT-CUP LIVERWORT

Lunularia cruciata

Easily recognized by the crescent-shaped gemma cups on the bright green thallus, this very common species occurs on river banks but is most frequent around human habitation – on damp paths and walls and in greenhouses, where it is often a weed in flower pots. Separate male and female plants. Spore capsules rarely formed.

CHAPTER FOUR

FUNGI

The fungi are a large group of plants which include the familiar mushrooms and toadstools as well as a host of lesser known forms – more than 50,000 species in all. The fungi are totally without chlorophyll and cannot, therefore, make their own food in the way that other plants do. In fact, some botanists do not consider the fungi to be plants at all. Most species live as saprophytes, obtaining food from dead and decaying matter – especially dead leaves on the forest floor – but there are also quite a number of parasitic species which attack living plants and animals. Examples include the rust fungi, which destroy vast amounts of cereals, and the fungi that cause athlete's foot and similar skin problems in human beings.

The fungus body consists essentially of a mass of slender threads called hyphae; the whole mass is known as the mycelium. The threads penetrate the food material and absorb nutrients from it; they are very often out of sight, but if you delve into the moist lower layers of the leaf litter on the woodland floor, you will discover just how extensive the mycelia are. They play a vital role in breaking down leaves and returning minerals to the soil.

Lower Fungi

All fungi reproduce by scattering spores – minute dust-like bodies that are normally carried away on the breeze. The classification of the fungi depends mainly on the way in which the spores are produced. The more primitive fungi, known as the Lower Fungi or, more scientifically, the Phycomycetes, bear their

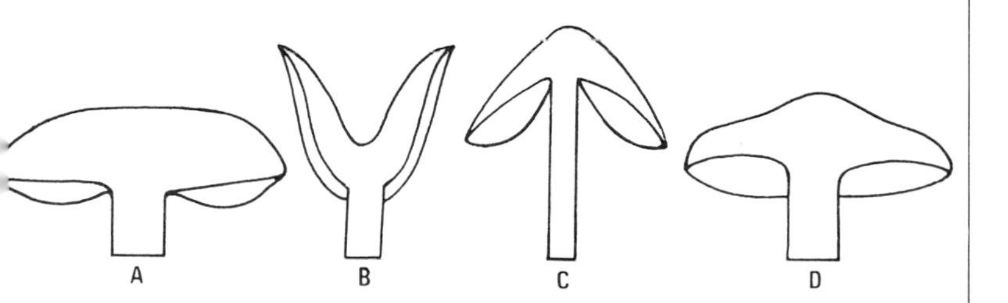

Fruitbody shape and gill attachment. A. shallow convex cap with free gills (not attached to stem). B. funnel-shaped cap with decurrent gills (running down stem). C. conical cap with adnexed gills (attached to stem by only part of the width). D. broadly campanulate cap with adnate gills (attached to stem by their complete width). Sinuate gills have a distinct notch close to the junction with the stem.

spores in small capsules on fairly simple threads. The Lower Fungi include many familiar moulds, such as the Pin Mould *(Mucor)* that often grows on stale bread and other foodstuffs.

Higher Fungi

With the exception of numerous mildews and yeasts and the *Penicillium* mould from which we get penicillin, the Higher Fungi produce their spores in relatively complex fruiting bodies. Formed of densely entwined threads, these are the only parts of the Higher Fungi that we normally see and they usually appear only at certain times of the year. Within the Higher Fungi two main groups are recognized: the Ascomycetes and the Basidiomycetes. The spores of the Ascomycetes develop inside small club-shaped or cylindrical cells called asci. These cells are normally carried at the surface of the fruiting body, the form of which is extremely varied (see pages 271-2). The spores of the basidiomycete fungi develop, usually in fours, on the outside of club-shaped cells called basidia; these structures are far too small to be seen without a microscope.

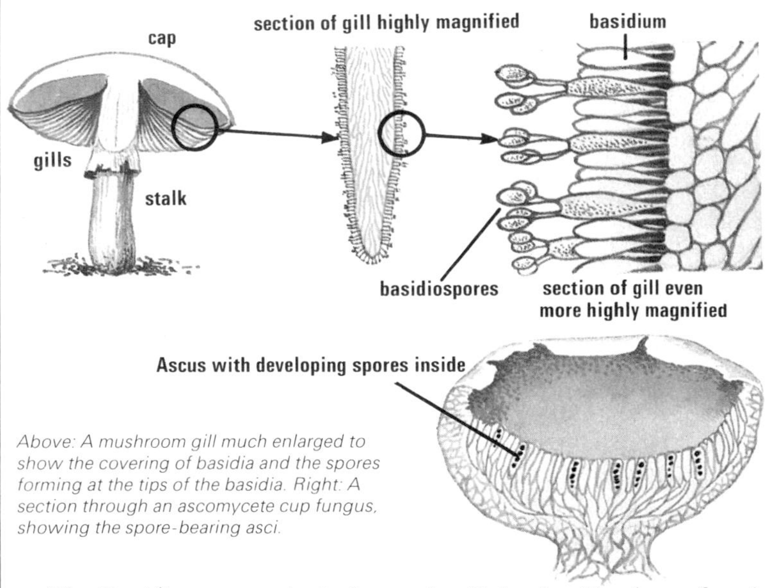

Above: A mushroom gill much enlarged to show the covering of basidia and the spores forming at the tips of the basidia. Right: A section through an ascomycete cup fungus, showing the spore-bearing asci.

The Basidiomycetes include nearly all the familiar large fungi; the following pages are devoted almost entirely to this important group. There are several divisions, of which the agarics are the best known. These are the typical mushrooms and toadstools, with a distinct cap on the underside of which there are numerous radiating gills. The term 'mushroom' was once restricted to the edible Field Mushroom *(Agaricus campestris)* and Horse Mushroom *(A. arvensis)* and their cultivated relatives. All other gill-bearing fungi were known as toadstools. Under American influence, however, all the agarics and most other large fungi now tend to be called mushrooms.

The basidia of the agarics clothe the gill surfaces and fire off the spores when they are ripe. The spores then fall and are carried away by the wind. By carefully separating a ripe cap from its stem and placing it on a sheet of paper, one can obtain a spore-print faithfully reproducing the pattern of the gills – this is often a help in identification. The spore colour is another useful identification guide, as is the way in which the gills are connected to the stem (see diagram). Some agarics exude a milky juice when the gills are broken; the nature of this 'milk' is also a good guide to the identity of the species (see page 243). Smell can even be used for distinguishing some species, while expert mycologists make use of a wide range of chemical tests.

Other major divisions of the Basidiomycetes include the boletes and the polypores, in which the underside of the cap is sponge-like with numerous minute pores. The basidia form the inner lining of the pores. These two groups can be separated quite easily because the boletes (see page 258) are normally of typical toadstool shape while the polypores generally form shelf-like outgrowths (brackets) on tree trunks and other wood (see pages 261-4). In addition, the pores of the boletes are easily stripped from the flesh forming the rest of the cap. The Gasteromycetes, represented by puff-balls and stinkhorns, are yet another division.

Edible or Poisonous?

Many fungi, including the Field Mushroom, the Horse Mushroom and most of the boletes are very good to eat, but most are worthless and some, such as the Death Cap *(Amanita phalloides)*, are deadly poisonous. Unfortunately, there is no simple rule for distinguishing edible from poisonous species and no fungus should be eaten before making a thorough check of its identity.

Lichens

Lichens (see page 273) are complex plants consisting of a fungus – usually an ascomycete – and an alga living in very close partnership (symbiosis). Each derives benefit from the partnership, but the fungus cannot survive alone. Although lichens are usually dry and quite unlike normal fungi, they are often regarded just as fungi that require the help of an alga to survive.

Basidiomycetes

GILL FUNGI (Agarics)

The agarics are those fleshy fungi which bear gills. They grow on the ground in pasture or woodland, on dung, trees or on woody debris. The stem may or may not bear a ring and may or may not show a sheathing, sac-like structure (the volva) at its base.

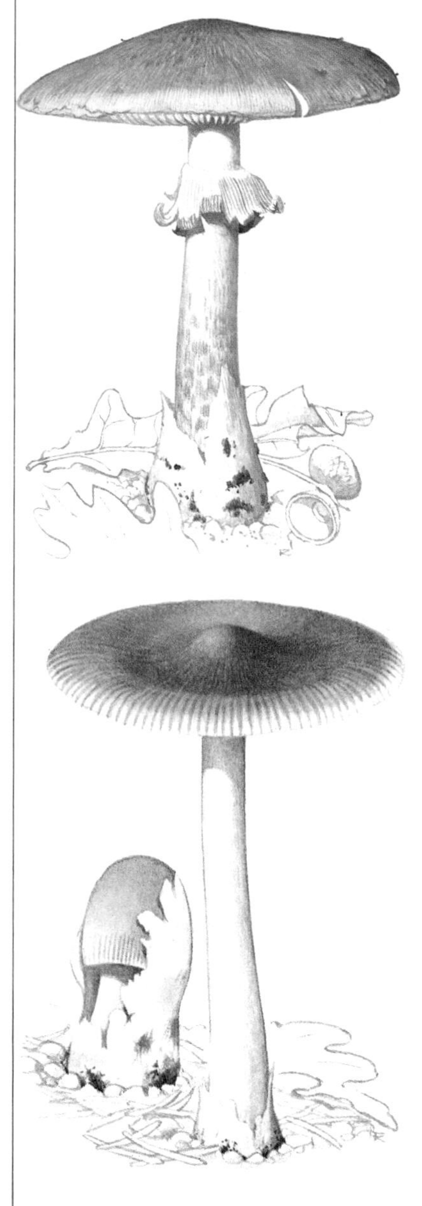

DEATH CAP
Amanita phalloides

Cap: 6-9 cm diam., convex then shallowly convex; varying in colour from olive-green at the centre to yellowish-green nearer the smooth margin. The cap surface appears indistinctly radially streaky; usually naked, without trace of warts or scales.
Stem: 7-9 cm high; 1-1·5 cm wide; white, cylindrical or narrowed above and sheathed below in a conspicuous free-standing, white, sac-like volva.
Ring: near apex of stem; white.
Gills: white; virtually free.
Smell: when old unpleasant, cheesy.
Spore-print: white.
Habitat: deciduous woodland, especially with beech and oak. Autumnal. Fairly common.
POISONOUS, often fatal.

TAWNY GRISETTE
Amanita fulva

Cap: 3·5-5 cm diam., long remaining acorn-shaped, then bell-shaped, finally flat but often with central boss, bright tawny-brown, often darker at centre, naked. Margin striate, fluted or grooved.
Stem: up to 11 cm high, 1 cm wide, tall, fragile, hollow, cylindrical, pale-tawny with well-developed, sac-shaped volva.
Ring: absent.
Gills: white, free.
Spore-print: white.
Habitat: coniferous and deciduous woodland. From late summer to autumn; one of the earliest mushrooms to appear. Very common. Edible.

FALSE DEATH CAP

Amanita citrina

Cap: 6-8 cm diam.; convex then flat; pale cream to pale lemon-yellow ornamented with a few large, or several smaller, thick, flat, whitish to brownish patches of velar tissue.
Stem: 8-11 cm high, 1-2 cm wide, tall in proportion to cap; white, cylindrical with a conspicuous basal bulb, up to 3 cm wide, with margin indicated by a prominent rim representing the volva.
Ring: near apex; white.
Gills: white to very pale cream.
Smell: of raw potato.
Spore-print: white.
Habitat: deciduous and coniferous woodland. Autumnal. Very common.

AMANITA PANTHERINA

Cap: 6-8 cm diam., convex, then flat, dark greyish-brown to olive-brown, sometimes paler and more yellowish-brown, ornamented with numerous, uniformly distributed small white pyramidal warts; margin somewhat striate.
Stem: 7-10 cm high, 1-1·5 cm wide, white, rather tall, cylindrical, slightly enlarged below where the volva disrupts into concentric rings, the uppermost forming a narrow, close-fitting but free collar or ridge.
Ring: near middle of stem, white.
Gills: white, free.
Flesh: white.
Spore-print: white.
Habitat: deciduous woodland. Autumnal. Rare. POISONOUS.

LEPIOTA CRISTATA

Cap: 2-4 cm diam., broadly campanulate, surface disrupting into tiny red-brown scales on a white background; the scales rapidly disappear except around a small, central, similarly coloured disc.
Stem: 2-5 cm high, 4-6 mm wide, white, cylindrical to slightly enlarged below.
Ring: near apex of stem, white, membranous, deciduous.
Gills: white, free.
Smell: unpleasant, sour.
Spore-print: white.
Habitat: deciduous woodland, often in grass along rides. Autumnal. Common. POISONOUS.

FLY AGARIC
Amanita muscaria

Cap: up to 15 cm diam., convex, flattened to saucer-shaped, scarlet ornamented with white warts which gradually disappear and may be completely lacking. The cap colour often fades to reddish-orange. In the button stage the cap is covered by thick white velar tissue which eventually cracks to form the scales and expose more and more of the red surface as expansion occurs.
Stem: up to 20 cm high, 3 cm wide, white to pale-yellowish above, cylindrical and brittle with a slightly broader base surmounted by a series of concentric zones of white scales representing the volva.
Ring: near apex of stem, white to very pale-yellowish.
Gills: white, almost free.
Habitat: Under birch, occasionally with pine and other trees. Autumnal. Common. POISONOUS.

BLUSHER
Amanita rubescens

Cap: 8-12 cm diam., at first strongly bell-shaped, then convex, finally flat to shallowly saucer-shaped, varying in colour from pale pinkish-brown with a darker red-brown centre to entirely red-brown, ornamented with thin mealy or hoary patches of whitish, greyish or pale volval tissue which tend to disappear with age.
Stem: 8-11 cm high, 2·5-3·5 cm wide, stout and stocky, cylindrical with somewhat enlarged base, whitish becoming pale pinkish-brown below, especially when handled. Volva reduced to inconspicuous rows of concentric scales.
Ring: near apex of stem; white.
Gills: white, often spotted red with age.
Flesh: white becoming pinkish when cut and in insect holes.
Spore-print: white.
Habitat: deciduous and coniferous woodland. Late summer and autumn. One of the first mushrooms to fruit. Very common. Edible, but best avoided owing to possible confusion with poisonous species.

PARASOL MUSHROOM
Lepiota procera

Cap: 12-22 cm diam., umbrella-shaped with nipple-like boss at centre, surface disrupting into large, often upturned, dark-brown scales on a dirty white, coarsely fibrillose background; scales larger and widely dispersed towards margin, smaller, more densely arranged at centre with nipple uniformly dark-brown.
Stem: 20-26 cm high, 1·5-2·0 cm wide, tall, narrow, cylindrical with bulbous base up to 4 cm wide, pale, ornamented with dark-brown, zig-zag markings.
Ring: large, spreading with double edge, whitish, movable.
Gills: soft to touch, free, attached to collar-like rim surrounding the stem apex.
Flesh: white.
Spore-print: white.
Habitat: pastures, edge of woods grassy rides. Autumnal. Occasional. Edible.

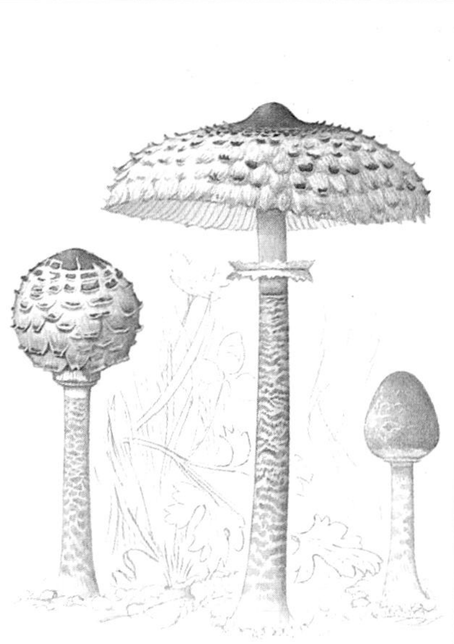

HONEY FUNGUS
Armillariella mellea

Cap: 6-12 cm diam., convex, then flat to saucer-shaped, tan, tawny or cinnamon-brown, paler towards the indistinctly striate margin; ornamented with delicate, dark-brown, hair-like scales, prominent and crowded in young specimens, eventually disappearing except at centre in old fruitbodies.
Stem : 9-14 cm high, 1-2 cm wide, tough, cylindrical, pale-tawny, whitish at apex.
Ring: near apex of stem, thick, cottony, whitish, often with yellow flocci at margin.
Gills: dirty-whitish to flesh-coloured; adnate.
Flesh: white, soft. Taste acrid.
Spore-print: cream.
Habitat: at base of living and dead trunks or stumps. Autumnal. Very common. Edible when young.

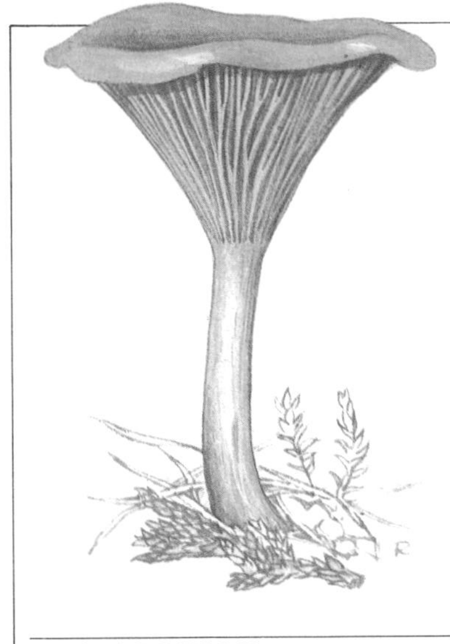

FALSE CHANTERELLE
Hygrophoropsis aurantiaca

Cap: 3-6 cm diam., funnel-shaped with enrolled margin, surface suede-like, orange or yellowish-orange.
Stem: 2-4 cm high, 5-7 mm wide, same colour as cap, often brownish below when old.
Gills: decurrent, crowded, repeatedly forked, deep orange.
Smell: not distinctive.
Spore-print: white.
Habitat: coniferous woodland and heaths. Autumnal. Common. Edible but worthless.

CHANTERELLE
Cantharellus cibarius

Cap: 2·5-6 cm diam., top-shaped, often slightly depressed at centre, gradually narrowed below into the short stalk; smooth, moist, entire fungus bright egg-yellow.
Stem: 2-6 cm high, 6-16 mm wide, short and squat.
Gills: decurrent, blunt, irregularly branched, fold-like and interconnected.
Smell: pleasant (of apricots).
Spore-print: white.
Habitat: deciduous woodland, especially on sandy or clay banks amongst moss. Autumnal. Occasional. Edible. Much sought after and collected for sale in markets. Easily dried for use in cooking. It may be confused with the False Chanterelle, but is much paler and altogether more fleshy.

LACTARIUS RUFUS

Cap: 2·5-4 cm diam., flat or slightly depressed, irregularly radially wrinkled about a small central nipple, orange-brown paling to yellowish-buff when dry, margin somewhat striate when moist.
Stem: 2-4 cm high, 4-6 mm wide, same colour as cap.
Gills: decurrent, buff.
Milk: mild, white, changing to yellow if allowed to dry on a handkerchief.
Spore-print: pale buff.
Habitat: deciduous woodland. Autumnal. Common.

UGLY ONE
Lactarius turpis

Cap: 4-6 cm diam., shallowly funnel-shaped with prominent central nipple, rich red-brown, surface smooth, dry, minutely grained, appearing falsely granular.
Stem: 4·5-5·5 cm high, 5-7 mm wide, same colour as cap but paler, base whitish.
Gills: decurrent, yellowish.
Milk: white, plentiful, very peppery, but only after a minute or so.
Spore-print: pale pinkish-buff.
Habitat: coniferous woodland, very rarely with birch. Autumnal. Common.

LACTARIUS TABIDUS

Cap: 8-20 cm diam., shallowly funnel-shaped or convex with depressed centre, surface sticky, smooth except for the enrolled felty margin, dark, dull olive-brown to almost black, brighter yellowish-olive towards edge.
Stem: 4·5-7 cm high, 2-2·5 cm wide; short, squat, sticky, same colour as, but paler than cap, often pitted.
Gills: more or less decurrent, dirty creamy-white becoming brown and discoloured when bruised.
Milk: white, plentiful, very peppery.
Spore-print: pale pinkish-buff.
Habitat: strictly associated with birch on heaths and commonland, often overgrown with grass. Autumn. Very common.

LACTARIUS DELICIOSUS

Cap: 6-10 cm diam., convex with depressed centre or shallowly funnel-shaped, moist, reddish-orange with darker greenish zones and variable development of green staining, surface appearing almost as if granular-stippled.
Stem: 6-8 cm high, 1·5-2·0 cm wide, orange, often pitted.
Gills: orange-yellow, staining greenish.
Milk: brilliant carrot-colour eventually becoming wine-red in 30 minutes.
Spore-print: pale pinkish-buff.
Habitat: coniferous woodland, especially pine. Autumnal. Common. Edible.

An unmistakable species due to its orange colour and green staining of all parts, and the presence of vivid carrot-coloured milk.

RUSSULA OCHROLEUCA

Cap: 5-8 cm diam., shallowly funnel-shaped, moist, bright ochre-yellow to greenish-yellow.
Stem: 6-8 cm high, 1·5-2 cm wide, rather soft with firmer rind, white, eventually greyish, surface ornamented with faint, densely-crowded short, raised longitudinal lines.
Gills: whitish.
Flesh: mild to moderately hot.
Spore-print: pale-cream.
Habitat: coniferous and deciduous woodland. Autumnal but persisting late in the season. Very common.

RUSSULA SARDONIA

Cap: 6-10 cm diam., convex to broadly bell-shaped, varying from dark reddish- or violet-purple to purplish-black.
Stem: 7-12 cm high, 1·5-2 cm wide, rather tall, beautifully purple.
Gills: adnexed, primrose, often with watery droplets along the edge.
Flesh: white, very hot.
Spore-print: cream.
Habitat: coniferous woodland. Autumnal. Common.

RUSSULA ATROPURPUREA

Cap: 4-10 cm diam., convex sometimes with slight hump, moist, dark reddish-purple to almost black, old specimens often mottled yellow at centre.
Stem: 5-7 cm high, 1·5-2 cm wide, short, squat, white, with rust-coloured base.
Gills: adnexed, whitish to pale cream, often discoloured with rusty spots.
Flesh: mild to slightly peppery.
Spore-print: white to off-white.
Habitat: deciduous woodland. Autumnal. Common.

SICKENER
Russula emetica

Cap: 4-7 cm diam., convex then depressed, brilliant scarlet with a moist shiny surface, margin eventually coarsely striate.
Stem: 5-8 cm high, 1·5-2 cm wide, rather tall, fragile, pure white, the lower portion usually somewhat club-shaped.
Gills: adnexed, white.
Flesh: very hot.
Spore-print: pure white.
Habitat: coniferous woodland. Autumnal. Common. May cause sickness if eaten raw.

RUSSULA CYANOXANTHA

Cap: 7-10 cm diam., convex, slightly depressed at centre, moist, dark greyish-purple with olive tones.
Stem: 8-11 cm high, 1·5-2 cm wide, white, hard.
Gills: white, softly pliable.
Flesh: mild.
Spore-print: white.
Habitat: deciduous woodland. Autumnal. Common.

POACHED EGG FUNGUS
Oudemansiella mucida

Cap: 3-7 cm diam., soft, flabby, convex, very glutinous, white becoming flushed with grey, margin striate, almost translucent.
Stem: 5-7 cm high, 4-6 mm wide, often curved, tough, cartilaginous, cylindrical; often expanded disc-like at point of attachment; white to greyish.
Ring: spreading, white above, greyish below especially toward edge.
Gills: distant; deep, soft, white.
Spore-print: white.
Habitat: confined to beech, occurring in small clusters on trunks, branches, or on stumps. Autumnal. Common.

CLOUDED CLITOCYBE
Clitocybe nebularis

Cap: 7-16 cm diam., robust, fleshy, convex with a low central hump to more or less flat, cloudy-grey sometimes with a brown tinge, surface dry appearing to have a faint hoary bloom.
Stem: 7-10 cm high, 1·5-2·5 cm wide, cylindrical, same colour as cap but paler, often striate.
Gills: decurrent, crowded, dirty creamy-white.
Smell: characteristic, unpleasant.
Spore-print: creamy-white.
Habitat: deciduous woodland, especially in areas rich in humus, near piles of rotting leaves or grass, usually gregarious and sometimes forming fairy-rings. Autumnal. Common. POISONOUS.

VELVET SHANK
Flammulina velutipes

Cap: 2·5-5 cm diam., shallowly convex becoming flat, bright-yellowish or orangey-tan, often darker and more brownish at the centre, moist becoming sticky when wet, shiny when dry.
Stem: 2·5-5 cm high, 4-6 mm wide, very dark-brown and conspicuously velvety, paling to yellow nearer the cap.
Gills: adnexed, rather distant, pale creamy-yellow.
Habitat: on trunks and branches, especially of dead elm, often forming small tiered clusters.

PLUMS AND CUSTARD

Tricholomopsis rutilans

Cap: 6-12 cm diam., convex to broadly bell-shaped, yellow, densely covered with tiny fleck-like purple scales which are continuous at centre but nearer the margin become pulled further apart to show more and more of the yellow background.
Stem: 6-9 cm high, 1-1·5 cm wide, similar in colour to cap, likewise densely flecked below with purple scales which disappear towards the yellow apex.
Gills: yellow.
Flesh: yellowish.
Spore-print: white.
Habitat: on conifer stumps. Autumnal. Fairly common.

BLEWIT

Lepista saeva

Cap: 6-8 cm diam., convex then flat, moist, varying from buff to greyish-buff.
Stem: 5-6 cm high, 1·5-2 cm wide, often enlarged at base, bright violet with streaky fibrillose surface.
Gills: whitish to pale flesh-coloured.
Spore-print: pale pinkish.
Habitat: in grassland, often forming fairy-rings. Autumnal. Uncommon. Edible and good, sometimes sold in shops.

WOOD BLEWIT

Lepista nuda

Cap: 6-10 cm diam., shallowly convex, becoming flat, often with greasy or water-soaked appearance, varying in colour from entirely violet to reddish-brown with violet tint localized to margin.
Stem: 6-8 cm high, 1·5-2 cm wide, bright bluish-lilac.
Gills: at first vivid violet becoming pinkish with age.
Spore-print: pale pinkish.
Habitat: deciduous woodland, compost heaps. Late autumn persisting into winter. Common. Edible and good, sometimes sold in shops.

DECEIVER
Laccaria laccata

Cap: 2-4 cm, shallowly convex to broadly bell-shaped, sometimes slightly depressed at centre; surface felty or scurfy; when moist bright red-brown with striate margin, drying out to pale buff and opaque and not striate.
Stem: 4-5 cm high, 5 mm wide, tough, streaky fibrillose, reddish-brown, often twisted.
Gills: deep, thick, distant, pinkish flesh-colour with waxy appearance.
Spore-print: white.
Habitat: deciduous woodland and coniferous woodland, heaths. Often gregarious. Autumnal. Very common.

Because of its variability in appearance due to the degree of moistness of the cap, it can be very difficult to recognize in all its guises, hence the common name. The thick, distant, waxy, flesh-coloured adnate or adnexed gills are a good guide to recognition.

AMETHYST DECEIVER
Laccaria amethystea

Cap: 2·5-4 cm diam., convex then shallowly convex, often depressed at centre, surface scurfy-felty, when moist entire fungus bright violet, but when dry this fades to pale buff with faint lilac tint.
Stem: 4-6 cm high, 6 mm wide, deep violet, but paler when dry.
Gills: deep, thick, distant, adnate or adnexed, bright violet fading to lilaceous-flesh-colour when dry.
Spore-print: white.
Habitat: deciduous woodland. Autumnal. Common.

SPINDLE SHANK
Collybia fusipes

Cap: 3-7 cm diam., broadly bell-shaped with central boss, smooth, dark red-brown drying pinkish buff or pale tan.
Stem: 8-10 cm high, 1-1·5 cm wide, tough, similarly coloured to cap, gradually enlarged below then conspicuously narrowed into a rooting portion, surface distinctly grooved.
Gills: adnexed, broad, distant, whitish then flushed with red-brown, also often spotted with brown.
Spore-print: white.
Habitat: tufted at base of tree trunks, especially oak. Late summer to early autumn. Occasional.

FOXY SPOT
Collybia maculata

Cap: 5-9 cm diam., convex, white with pinhead-sized or larger red-brown spots, often becoming entirely pale red-brown with age.
Stem: 8-10 cm high, 1-1·6 cm wide, firm, white, often longitudinally striate, tapering below into a short rooting base.
Gills: densely crowded, shallow, pale-cream spotted with red-brown.
Habitat: coniferous woodland or amongst bracken in heathy situations where it sometimes forms fairy-rings. Autumnal. Common.

WOOD WOOLLY FOOT
Collybia peronata

Cap: 4-6 cm diam., very broadly bell-shaped to flat, sometimes with central boss, smooth, very tough and leathery, ochre-coloured to reddish-brown, drying paler.
Stem: 7-9 cm high, 5 mm wide, tall, narrow, but very tough, pale yellowish-buff, thickly covered toward base with pale yellowish woolliness.
Gills: tough, leathery, distant, adnexed, separating from around the top of the stem in a false collar; same colour as cap.
Flesh: thin, leathery, yellowish, taste peppery.
Spore-print: white.
Habitat: especially deciduous woodland amongst leaf litter. Autumnal. Common.

FAIRY-RING CHAMPIGNON
Marasmius oreades

Cap: 3-5 cm diam., convex with broad central boss, tough, smooth, pinkish-tan drying to pale buff, margin often grooved.
Stem: 4·5-5·5 cm high, 2·5-3 mm wide, firm, tough, pale buff.
Gills: adnexed, deep, distant, whitish.
Spore-print: white.
Habitat: pastures, lawns, roadside verges and the commonest cause of 'fairy-rings' in turf. Late summer to autumn. Common. Edible.

Recognized when forming fairy-rings by the tough, broadly bell-shaped, buff-coloured cap with distant gills. However it does not always grow in rings.

ST GEORGE'S MUSHROOM
Tricholoma gambosum

Cap: 5-10 cm diam., shallowly convex, often with undulating margin, smooth, white to very pale buff at centre.
Stem: 4-7 cm high, 1·5-2 cm wide, short, squat, white.
Gills: sinuate, crowded, white.
Flesh: thick, smelling strongly of meal.
Spore-print: white.
Habitat: in pastures, roadside verges, hedge-bottoms. This whitish mushroom is readily recognized by its squat, fleshy fruitbodies which occur in spring, usually around St. George's Day (April 23rd), hence the common name. Edible.

ENTOLOMA CLYPEATUM

Cap: 3-6 cm diam., bell-shaped, then shallowly bell-shaped with central boss, grey-brown with darker radial streaks, drying paler.
Stem: 4-6 cm high, 6-15 mm wide, dirty-white to greyish with fibrillose surface.
Gills: sinuate, deep, distant, greyish becoming pink, edge irregularly wavy.
Flesh: greyish when water-soaked, white when dry, smelling of meal when crushed.
Spore-print: pink.
Habitat: often associated with rosaceous trees and shrubs, such as hawthorn, and common in hedgerows. Also in rich garden soil. Vernal.

HYGROPHORUS CONICUS

Cap: up to 3 cm high, acutely conical with fibrillose surface, yellow or orange becoming black on handling or with age.
Stem: up to 6 cm high, 5 mm wide, fibrillose, yellow, then blackening.
Gills: ascending, almost free, pale yellow.
Spore-print: white.
Habitat: in grassland. Autumnal. Fairly common.

The acutely conical fibrillose orange cap which, like the rest of the fruit-body blackens when handled, is very distinctive.

MILKING MYCENA

Mycena galopus

Cap: 1 cm diam., broadly bell-shaped, pale with brown centre and radiating striae.
Stem: up to 5 cm high, 2 mm wide, greyish below, almost white above, when broken exuding a white milk.
Gills: adnexed, white.
Spore-print: white.
Habitat: woodland, hedge-bottoms, heaths etc. Autumnal. Very common.

The pale cap with brown radiating striae and presence of a white milk in the stem are distinctive characters, although it may not be possible to obtain milk from the stems of old fruiting bodies.

LEATHERY MYCENA

Mycena galericulata

Cap: 2-4·5 cm diam., broadly bell-shaped, flat with a central boss, tough, leathery, varying from grey-brown to buff, margin striate to somewhat grooved.
Stem: 7-10 cm high, 3-5 mm wide, tough, cartilaginous, smooth, polished, grey-brown.
Gills: adnate with decurrent tooth, deep, distant, interveined, whitish then flesh-coloured.
Smell: mealy when crushed.
Spore-print: white.
Habitat: clustered on stumps, or from buried wood. Typically autumnal, but occurring sporadically throughout the year. Very common.

SHAGGY PHOLIOTA

Pholiota squarrosa

Cap: 6-8 cm diam., convex, dry, pale ochre, entirely covered with prominent densely crowded upturned bristly scales giving a coarse shaggy appearance.
Stem: 6-10 cm high, 1-1·5 cm wide, same colour as cap and covered with similar recurved scales below the fibrillose ring.
Gills: adnate with decurrent tooth, crowded, at first yellowish then pale rust-coloured.
Flesh: pale.
Spore-print: rusty-brown.
Habitat: parasitic on many species of deciduous trees, forming dense shaggy tufts at the base of living trunks. Autumnal. Fairly common.

OYSTER FUNGUS
Pleurotus ostreatus

Cap: 5-12 cm wide, tiered, convex, bracket-shaped, smooth; variable in colour from blue-grey to buff when young, becoming dark brown with age.
Stem: 2-3 cm long, 1·5-2 cm wide, short, thick, lateral, hairy, white.
Gills: decurrent, distant, whitish.
Spore-print: lilac.
Habitat: in tiered clusters on standing or fallen trunks especially beech. Autumnal but found occasionally at other times of year. Common. Edible.

GYMNOPILUS JUNONIUS

Cap: 6-12 cm diam., convex, fleshy, bright tawny or golden yellow; surface fibrillose or disrupting into indistinct fibrillose scales.
Stem: 7-15 cm high, 1·2-3 cm wide, swollen near middle and then tapering towards base, fibrillose, same colour as cap, but paler, and with membranous ring near apex.
Ring: yellowish, soon collapsing back onto stem.
Gills: adnate, crowded, sometimes with decurrent tooth, yellow becoming rust-coloured.
Flesh: pale yellowish.
Spore-print: rusty-brown.
Habitat: forming dense tufts at the base of trunks (sometimes living) or on stumps of deciduous trees. Autumnal. Common.

PSATHYRELLA GRACILIS

Cap: 1·5-2·5 cm diam., bell-shaped, dark brown or reddish-brown with striate margin when moist, opaque and pale biscuit colour when dry.
Stem: 8-10 cm high, 2 mm wide, very tall, fragile, pure white with white hairy fibrils at base.
Gills: adnate, blackish, with pink edge.
Habitat: solitary or gregarious, amongst grass or leaves, in deciduous woodland, hedge-bottoms, roadside verges. Autumn. Very common.

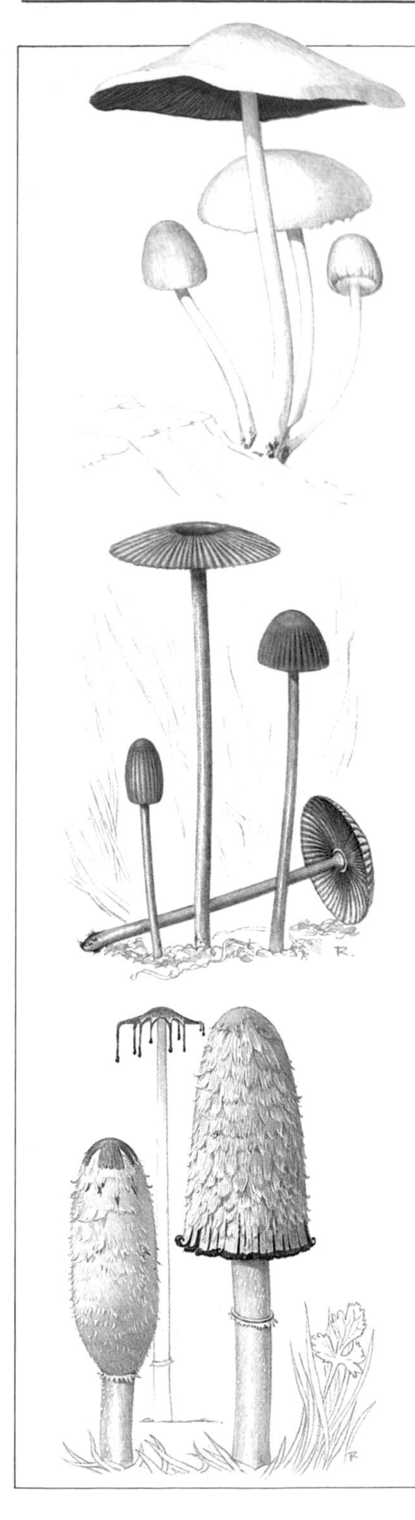

PSATHYRELLA CANDOLLEANA

Cap: 2·5-5 cm diam., shallowly bell-shaped to flat, pale creamy-ochre to whitish especially when dry, with tiny tooth-like remnants of veil hanging from margin when young.
Stem: 4-6 cm high, 4-5 mm wide, white, hollow, very brittle.
Gills: adnate, for a long time whitish then lilac-grey and finally brownish-black.
Spore-print: almost black.
Habitat: tufted on wood, stumps, roots, fence-posts. Spring to Autumn. Common.

LITTLE JAPANESE UMBRELLA
Coprinus plicatilis

Cap: 2-3 cm diam., acorn-shaped, yellowish-brown and closely striate at first, then flat, coarsely grooved or fluted, grey with small depressed tan-coloured disc, very thin, short-lived, almost translucent.
Stem: 6-8 cm high, 3 mm wide, whitish, very delicate.
Gills: free, attached to a collar around stem apex, scarcely liquefying.
Spore-print: black.
Habitat: damp grass, lawns, road-side verges. Autumnal. Common.

SHAGGY INK CAP or LAWYER'S WIG
Coprinus comatus

Cap: 6-14 cm high, cylindrical, opening slightly at base, eventually bell-shaped, white, with buff-coloured central disc, surface broken up into shaggy scales, margin closely striate becoming greyish, finally black on maturity. Entire cap gradually dripping away from margin as an inky fluid. This fluid contains the spores and has been used as ink.
Stem: up to 30 cm high, 1-1·5 cm wide, white with movable membranous ring toward base.
Gills: free, crowded, white near stem apex, then pink and finally black at margin.
Spore-print: black.
Habitat: gregarious on rubbish tips, roadside verges, fields and gardens. Autumnal. Fairly common. Edible, provided the gills have not started to liquefy.

ANTABUSE INK CAP
Coprinus atramentarius

Cap: 5-7 cm high, bell-shaped, irregularly ribbed and wrinkled almost to disc, grey, often with few inconspicuous brownish scales at centre.
Stem: 7-9 cm high, about 1 cm wide at base, where there is a conspicuous oblique ring-zone, white.
Gills: free, whitish, becoming grey, finally black, liquefying.
Spore-print: black.
Habitat: tufted in vicinity of stumps of deciduous trees or from roots, often in gardens, fields etc. Autumnal but sporadically throughout the year from early spring. Common. Edible, but causing sickness if eaten with alcohol.

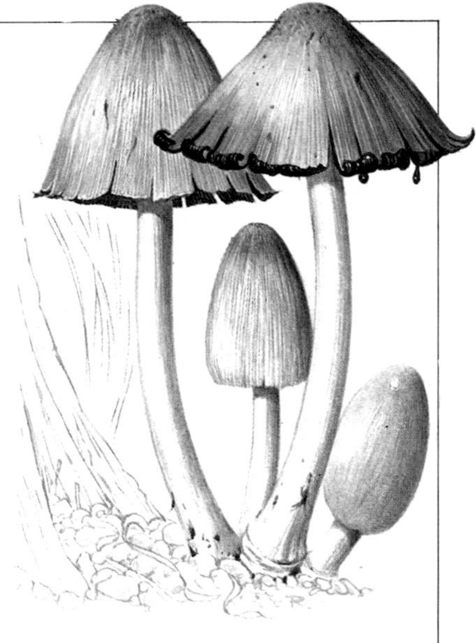

TROOPING CRUMBLE CAP
Coprinus disseminatus

Cap: 5-10 mm high, acorn-shaped to hemispherical, pale yellowish-clay becoming greyish at the margin, closely grooved almost to the centre, minutely hairy under a very strong lens.
Stem: 1-3·5 cm high, 1-1·5 mm wide, white, very delicate and brittle.
Gills: adnate, dark grey to blackish, scarcely liquefying.
Spore-print: black.
Habitat: densely gregarious, covering entire stumps in myriads of tiny, brittle, bell-shaped, biscuit-coloured fruitbodies. Autumnal, but also sporadically throughout the year. Common.

FIELD MUSHROOM
Agaricus campestris

Cap: 4·5-8 cm diam., convex then flat, white, surface often disrupting into indistinct fibrillose scales especially around centre.
Stem: 4-6 cm high, 1-1·5 cm wide, short, squat, with pointed base and ring.
Ring: poorly developed, simple, often little more than a torn fringe.
Gills: free, at first pink then purplish-brown.
Flesh: white, sometimes reddish in stem when cut.
Spore-print: purplish-brown.
Habitat: grassy places, pastures, often in fairy-rings. Autumnal. Occasional. Edible.

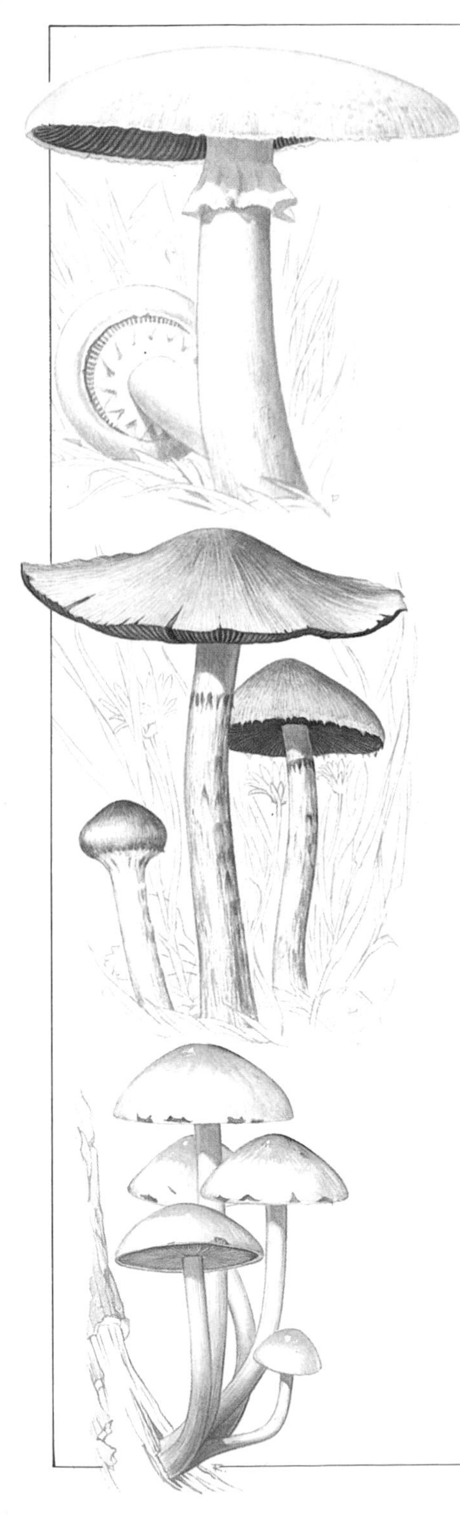

HORSE MUSHROOM
Agaricus arvensis

Cap: 6-11 cm diam., hemispherical, becoming shallowly convex, white, often creamy with age, sometimes faintly yellow when handled, but never vividly so.
Stem: 8-12 cm high, 1·5-2 cm wide, tall, cylindrical, with bulbous base, white with membranous ring.
Ring: large, pendulous, high on stem, underside like radiating spokes of a wheel (cog-wheel-like).
Gills: white, then brownish, finally purplish-brown, never pink.
Flesh: white, unchanging.
Spore-print: purplish-brown.
Habitat: open grassland, hillsides, orchards etc, often in fairy-rings. Autumnal. Occasional. Edible.

WEEPING WIDOW
Lacrymaria velutina

Cap: 4-6 cm diam., bell-shaped or convex, yellowish-brown to clay-brown, densely radially fibrillose, with enrolled woolly fringed margin.
Stem: 6-7 cm high, 5-8 mm wide, same colour as, but paler than cap, fibrillose or scaly, with prominent ring-zone of whitish cottony fibrils often becoming black due to trapped spores.
Gills: adnexed or adnate, almost black, mottled, with white edge bearing watery droplets in damp weather.
Spore-print: almost black.
Habitat: tufted from roots or buried wood, in deciduous woodland. Spring to autumn. Common.

SULPHUR TUFT
Hypholoma fasciculare

Cap: 2-4 cm diam., shallowly bell-shaped to shallowly convex, sulphur-yellow often with tawny flush at centre, margin with dark fibrillose remnants of veil.
Stem: 4-7 cm high, 4-8 mm wide, same colour as cap, sometimes flushed brown below, with poorly developed purplish-brown fibrillose ring-zone near apex.
Gills: sinuate, sulphur-yellow becoming olive.
Flesh: yellow with bitter taste.
Spore-print: purplish-brown.
Habitat: tufted at base of deciduous and coniferous tree-stumps. Autumnal, but sporadically throughout the year. Very common.

GALERINA MUTABILIS

Cap: 2-3·5 cm diam., convex to broadly bell-shaped, watery date brown with striate margin when moist but drying out conspicuously from centre to tan colour. Fruitbodies when semi-dry are sharply two-coloured with tan centre and an abrupt, broad, watery-brown marginal zone.
Stem: 3-5·5 cm high, 3·5-5 mm wide, pale yellowish above becoming dark-brown and covered with recurved scales below the ring.
Gills: adnate with decurrent tooth, pale then cinnamon-brown.
Spore-print: cinnamon-brown.
Habitat: tufted on stumps of broadleaved trees. Autumnal. Fairly common. Edible.

ROLL RIM
Paxillus involutus

Cap: 5-11 cm diam., convex with strongly enrolled margin, becoming shallowly depressed at centre, surface glutinous in wet weather, yellowish-brown, smooth except toward the edge which is often ribbed and downy.
Stem: 6-7 cm high, 1-1·2 cm wide, central, same colour as cap but paler, often streaky.
Gills: decurrent, yellowish-brown becoming dark red-brown when bruised.
Spore-print: ochre-brown.
Habitat: heathy places, associated with birch. Autumnal. Very common. POISONOUS.

AGROCYBE PRAECOX

Cap: 3-5 cm diam., convex, smooth, cream-coloured with ochre-coloured flush at centre.
Stem: 5-7 cm high, 5-7 mm wide, tall, slender, white, with ring.
Ring: membranous, whitish.
Gills: adnate, clay-brown, crowded.
Flesh: white in cap, brown in stem, smelling of meal when crushed.
Spore-print: clay brown.
Habitat: in grassy places, roadside verges. Vernal. Common.

BOLETES

Fruitbodies are fleshy, with a cap and central stalk which may or may not bear a ring. The underside of the cap is sponge-like with tiny pores. These are the openings of densely crowded tubes lined with basidia which produce the spores.

BOLETUS BADIUS

Cap: 6-10 cm diam., shallowly convex, dark bay to chocolate-brown, slightly sticky in wet weather, shiny when dry but softly downy at margin.
Stem: 7-8 cm high, 1·5-2 cm wide, pale brown with darker streaks.
Pores: adnate or adnexed, small, cream to lemon-yellowish, turning blue-green when bruised.
Habitat: deciduous and coniferous woodland. Autumnal. Common. Edible.

BOLETUS CHRYSENTERON

Cap: 5-7 cm diam., convex, madder-brown sometimes with olive tint, cracking in a chequered manner showing pale pinkish flesh in between; surface minutely felty.
Stem: 6-8 cm high, 1 cm wide, yellowish streaked red below.
Pores: adnate to adnexed, large angular, at first pale dull yellow then olive-yellow.
Flesh: pink under cuticle, yellowish elsewhere often reddish in stem, turning slightly blue when cut.
Habitat: deciduous woodland. Autumnal. Very common.
B. subtomentosus, a common species of heathy areas, differs from *B. chrysenteron* in its velvety olive-tan coloured cap which bruises dark brown on handling. Its cuticle also seldom cracks and it lacks the pink colour beneath it.

BOLETUS ERYTHROPUS

Cap: 7-11 cm diam., convex, dry, minutely downy, dark bay-brown to red-brown.
Stem: 8-11 cm high, 1·5-2·5 cm wide, yellow densely covered with very minute red granular-dots.
Pores: free, minute, blood-red.
Flesh: yellow, instantly indigo-blue when cut or bruised.
Habitat: deciduous woodland. Autumnal. Fairly common. Edible despite startling colour change of flesh.

PENNY BUN BOLETE
Boletus edulis

Cap: 10-16 cm diam., strongly convex, usually chestnut brown.
Stem: 6-12 cm high, 2·5-4 cm wide, cylindrical but sometimes conspicuously swollen at base to 10 cm wide; pale with whitish network at least at apex.
Pores: adnexed, whitish becoming greenish-yellow.
Flesh: whitish unchanging, sometimes faintly pink in cap.
Habitat: deciduous and coniferous woodland. Late summer to autumn. Fairly common. Edible and excellent. (The 'Cep' or 'Steinpilz' of continental gourmets and a major constituent of many 'mushroom' soups).

DEVIL'S BOLETUS
Boletus satanas

Cap: 10-18 cm diam., convex with enrolled margin, pale-greyish.
Stem: 7-12 cm high, enormously swollen below up to 12 cm wide, yellow above, red below with conspicuous red network.
Pores: free, minute, blood-red.
Flesh: pale yellow, turns faintly blue in stem apex and over tubes.
Habitat: beechwoods on chalk. Autumnal. Rare. POISONOUS.

The large size, pale grey cap, enormously swollen stem with red net and the red pores are the salient features.

JERSEY COW BOLETE
Suillus bovinus

Cap: 4-7 cm diam., convex at first then flattened, glutinous when moist, buff to pinkish-buff with pale margin.
Stem: 4·5-7 cm high, 6-9 mm wide, tapering downward, same colour as cap.
Ring: lacking.
Pores: decurrent, large, irregular, compound (each pore subdivided into smaller pores), dirty-yellow to rusty.
Flesh: yellowish or pinkish, reddish in stem.
Habitat: coniferous woodland. Autumnal. Common. Edible.

Recognized by its colour (resembling that of a Jersey cow) and large compound pores.

SUILLUS LUTEUS

Cap: 7-12 cm diam., bell-shaped, then flattened, often with slight central boss, glutinous when moist, dark chocolate or purplish-brown, sometimes becoming rusty-tan with age.

Stem: 6-8 cm high, 1·7-2 cm wide, yellow with darker glandular dots above the well developed ring, whitish or pale brownish below.

Ring: spreading, membranous, white or greyish, often dark with age.

Pores: adnate to decurrent, dull yellow to deep ochre-yellow.

Flesh: whitish to pale lemon-yellow especially in stem.

Habitat: coniferous woodland. Autumnal. Fairly common. Edible.

LECCINUM SCABRUM

Cap: 5-10 cm diam., convex to shallowly convex, grey-brown with an almost granular-mottled effect under a lens.

Stem: 8-12 cm high, 2-3 cm wide, tall, cylindrical, often enlarged below, white but rough with conspicuous black flocci.

Pores: almost free, minute, dingy buff, bruising yellowish-brown.

Flesh: white, unchanging or faintly pink.

Habitat: with birch, especially on heaths. Autumnal. Very common. Edible.

POLYPORES

These are the bracket fungi and are recognized by the poroid undersurface. The spores are produced by the basidia which line the tubes, the openings of which form the pores. The fungi are mostly lignicolous (growing on wood). Most fruitbodies are laterally stalked or in the form of solitary or tiered bracket-like structures, commonly with a woody or leathery texture. Some form rosette-like clusters.

COLTRICIA PERENNIS

Cap: 2-10 cm diam., thin, leathery, funnel-shaped, adjacent fruitbodies often fused together, surface velvety, zoned in shades of tawny or rusty-brown but sometimes paling out to greyish-buff toward centre.
Stem: 3-5·5 cm high, 3-6 mm wide, velvety, rusty-brown.
Pores: small; lighter, brighter and more yellowish than stem, except when old, often with silky sheen.
Flesh: thin, fibrous, brown.
Spore-print: pale ochre.
Habitat: sandy heaths, sometimes on burnt ground. Autumnal but persisting for many months as discoloured blackened specimens.

GRIFOLA FRONDOSA

Fruitbody: up to 30 cm diam., consisting of a mass of small, thin, smoky-brown, often zoned, fan-shaped lobes, 3-7 cm across, each narrowed behind into a white stem-like portion attached to a common base.
Pores: decurrent, irregular, white.
Flesh: thin, white.
Spore-print: white.
Habitat: at base of living or dead stumps and trunks of deciduous trees, especially oaks. Summer to autumn. Occasional. Edible, but tough and with a mouse-like smell.

DRYAD'S SADDLE
Polyporus squamosus

Cap: 13-50 cm diam., fan-shaped, pale fawn with concentric rings of brown scales.
Stem: 3-8 cm high, 3-8 cm wide, relatively short, pale above with network due to rudimentary decurrent pores, black and swollen below.
Pores: white to cream, very large, 1-3 mm diam.
Flesh: white, rubbery, up to 4 cm thick behind, smell mealy.
Spore-print: white.
Habitat: parasitic on trunks of deciduous trees fruiting on the living or dead host, particularly common on elm, beech and sycamore, often at considerable height above ground. Spring to summer. Frequent. Edible but worthless.

GANODERMA APPLANATUM

Fruitbody: up to 30 cm diam. or more; flat, sessile, bracket-like, perennial, with thick, horny, dull crust, surface irregularly undulating, ornamented with conspicuous concentric grooves and varying in colour from buff to fawn or cocoa-brown; margin thin and rather acute.
Pores: very small, whitish, bruising brown.
Tubes: brown.
Flesh: fibrous, cinnamon-brown, thinner than tube layers.
Texture: very hard and woody.
Spore-print: cocoa-brown.
Habitat: parasitic on trunks and stumps, especially beech. Perennial. Very common.

BIRCH BRACKET
Piptoporus betulinus

Fruitbody: up to 20 cm diam., hoof-shaped, occasionally with a basal boss forming a short pseudo-stem, surface covered by a greyish to pale-brown or brown separable smooth skin; margin rounded.
Pores: very small, late-forming, white.
Flesh: white, rubbery, up to 7 cm thick behind.
Habitat: parasitic on birch, fruiting on the living and dead trunks as individual brackets although several may be present at intervals up the same tree. All the year round. Very common.

CORIOLUS VERSICOLOR

Fruitbodies: 3-8 cm diam., sometimes more when adjacent specimens merge together; thin, flexible sessile bracket-like, with felty surface ornamented with numerous colour zones varying from almost black or blue-black through smoky brown to tan or fawn especially at the margin. In addition to these colour zones there are often dark shining bands devoid of felt where the underlying surface shows through.
Pores: small, pale cream.
Flesh: white, about 2-3 mm thick.
Habitat: often conspicuously tiered, sometimes solitary, on stumps, trunks, fallen branches, bean poles or timber of both deciduous and coniferous trees throughout the year. Very common.

DAEDALEOPSIS CONFRAGOSA

Fruitbodies: up to 15 cm diam., flattened, shell-shaped, sometimes with a thickened basal hump at point of attachment, surface concentrically grooved and often irregularly radially wrinkled, zoned in lighter or darker shades of red-brown; margin acute, often white.
Pores: radially elongated, slot-like, white to faintly grey, bruising red when rubbed if fruitbody is in active growth, and becoming lilac with a drop of ammonia; on old fruitbodies the pores become uniformly reddish-brown. Spore-print white.
Flesh: rubbery, zoned, whitish then reddish- or pale-brown.
Habitat: solitary or gregarious, several brackets often produced at intervals up a trunk or along branches of deciduous trees, especially willow. Throughout the year. Common.

The brackets usually occur on small trunks or branches and frequently appear to grasp them or encompass them.

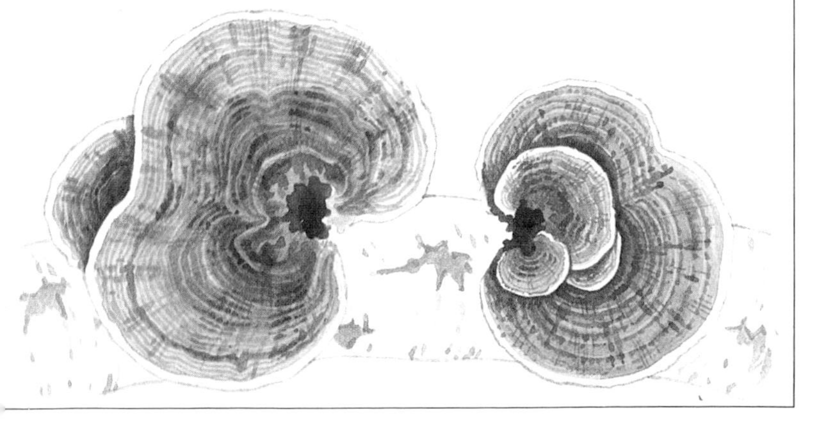

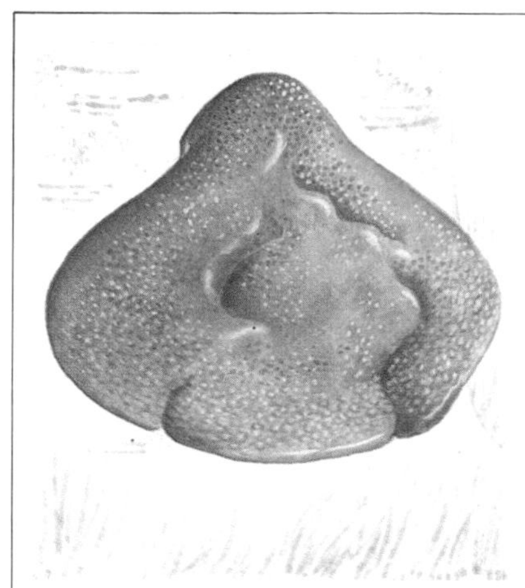

BEEFSTEAK FUNGUS
Fistulina hepatica

Fruitbodies: up to 15 cm diam., solitary, fan-shaped with a narrow point of attachment, sometimes with a short stalk-like base, liver-coloured, the surface roughened toward the margin with minute warts which are rudimentary tubes, otherwise smooth.
Tubes: individually separate.
Pores: yellowish-flesh-coloured.
Flesh: up to 5 cm thick behind with colour, graining and texture of raw beef-steak, also exuding a red juice; strong acidic taste.
Habitat: on oak stumps or trunks, usually near the ground. Autumnal. Occasional. Edible but of poor quality.

STEREOID AND THELEPHOROID FUNGI

Fruitbodies either terrestrial or lignicolous (growing on wood). Variable in shape: bracket-like, rosette-like, or coral-like with flattened branches. Fertile surface either smooth or wrinkled, but devoid of gills, pores or spines.

STEREUM RUGOSUM

Fruitbodies: forming extensive creamy flat patches on undersides of branches, only occasionally developing a very narrow, rigid, shelf-like portion, with dark-brown surface which may be slightly felty or naked. When scratched the cream-coloured surface reddens, and if broken the fruitbody is seen to have a stratified appearance under a lens.
Habitat: on small trunks, undersides of dead fallen or still attached branches of deciduous trees, especially hazel. All the year round. Common.

The creamy patches which redden when scratched are distinctive but if the fruitbody is dry when collected it may be necessary to moisten it before the red colour will develop.

HYMENOCHAETE RUBIGINOSA

Fruitbodies: 4-7 cm diam., densely-tiered, bracket-like, concentrically ridged and grooved, dark-brown to almost black and smooth, but when young with a minute rusty, velvety bloom. Lower surface chocolate-brown appearing as if waxed, with the extreme margin rust-brown or sometimes creamy-yellow. Flesh brown.
Habitat: old oak stumps. All the year round. Common.

The densely crowded dark brown to blackish brackets on oak stumps are easily identified, especially as they are rigid and brittle.

THELEPHORA TERRESTRIS

Fruitbodies: 2-6 cm diam., forming small irregular rosettes, either ascending or closely pressed to the ground, dark chocolate-brown, with soft, spongy, felt-like surface marked with radiating fibrils. Undersurface somewhat wrinkled and ornamented with minute warts, cocoa-brown.
Habitat: on the ground in coniferous woodland or on open sandy heaths. Autumnal. Common.

The small brown rosettes with soft felty surface are characteristic.

SILVER LEAF FUNGUS
Chondrostereum purpureum

Fruitbodies: 2-4 cm diam., thin, flexible leathery brackets, often arising from a flat area, sometimes forming entirely flat patches. Upper surface felty with one or more concentric grooves, pale greyish-buff, often with dark line at or just in from the wavy margin. Undersurface bright lilac to purplish when in active growth, fading to brownish with age.
Habitat: saprophytic on a wide range of deciduous trees (stumps, branches etc) or parasitic especially on rosaceous trees and shrubs. When parasitic it sometimes causes silvering of the foliage as in Silver Leaf Disease of plum. All the year round. Common.

JELLY FUNGI

Fruit bodies mostly lignicolous, growing on either living or dead wood: very variable in shape, but nearly always distinctly gelatinous and often brightly coloured when wet. They shrink a good deal if they become dry, assuming a rigid and horny texture and often becoming rather inconspicuous. Some dry right up to a varnish-like patch, but all regain their normal gelatinous and colourful state when they become wet again. Most of the 500 or so known species live in the tropics, where they run little risk of drying out.

JEW'S EAR
Auricularia auricula-judae

Fruitbodies: 3-6 cm diam., helmet-shaped, date-brown, firm-gelatinous or distinctly rubbery. Outer surface velvety. Undersurface, which bears the spores, has folds and ridges resembling the inside of an ear, pale purplish-brown; shiny.
Habitat: on branches of deciduous trees, especially elder. All the year round. Very common. Edible.

Distinctive because of its shape, colour and gelatinous texture. A white form also occurs and should be reported if found.

TREMELLA MESENTERICA

Fruitbodies: 1·5-5 cm diam., comprising pendant yellow gelatinous brain-like masses.
Habitat: on dead fallen or still attached branches of deciduous trees and shrubs. All the year. Common.

CALOCERA VISCOSA

Fruitbodies: 4-8 cm high, branched, club-shaped, bright egg-yellow, but of tough gelatinous texture; branches united below into a white rooting portion.
Habitat: on rotting conifer stumps. Autumnal. Common.

The bright yellow, branched, club-shaped fruitbodies on conifer stumps are distinctive. They are distinguished from the clavarias by their tough gelatinous texture, gliding easily between the fingers without breaking. The clavarias, commonly known as fairy clubs, are simple club-like or spiky fungi with very brittle flesh. Most are white, greyish, or yellow.

GASTEROMYCETES

The fungi in this group include puff-balls, earth stars (*Geastrum* species), stinkhorns, and bird's nest fungi. All produce their spores *inside* the fruiting body, which does not open until the spores are ripe. The Stinkhorn starts out as an egg-shaped body with a thick gelatinous wall, through which emerges the stout stalk and bell-shaped cap covered by the evil-smelling spore mass. Puff-balls and earth stars split open in various ways and the spores are puffed out when the fungus is disturbed. A single raindrop is enough to send out a cloud of spores, which are further dispersed by the breeze.

GEASTRUM STRIATUM

Fruitbodies: 3-6 cm diam., when expanded.
Spore-sac: about 1 cm diam., shortly stalked, flattened or sub-globose and flattened below, with a distinct rim or collar at base; bursts open through well-defined pore at top of a fluted cone, colour becoming almost blackish when old.
Rays: up to 8, becoming curled under fruitbody and lifting it clear off the ground.
Habitat: deciduous woodland. Autumnal. Occasional.

GEASTRUM TRIPLEX

Fruitbodies: 6-10 cm diam., when expanded.
Spore-sac: up to 3 cm diam., smooth, sessile, globular, pale-brown, seated in a shallow cup.
Rays: 4-8, thick, fleshy, yellowish to pale brown and curling under the spore sac and its cup.
Habitat: deciduous woodland. Autumnal. Occasional.

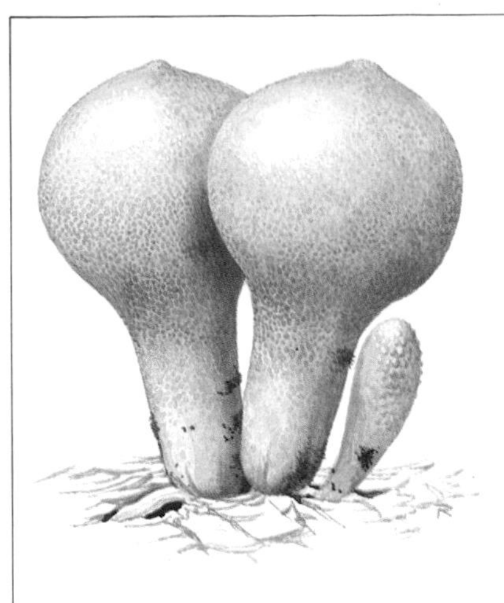

LYCOPERDON PERLATUM

Fruitbodies: up to 8 cm high, and 5 cm diam., club-shaped with a distinct head and long cylindrical stem, white to pale-brown, densely covered above with prominent, but deciduous, crowded, white pyramidal warts each surrounded by a ring of minute granules. In old specimens from which the warts have disappeared the surface has a reticulate pattern formed by the minute rings of granules.
Stalk: sponge-like.
Spore-mass: olive-brown.
Habitat: solitary, gregarious or even in small tufts in woodland. Autumnal. Very common.

BOVISTA NIGRESCENS

Fruitbodies: 3-6 cm diam., globular, at first white, but outer surface flaking away completely at maturity to expose the somewhat shiny blackish tissue beneath.
Stalk: lacking; fruitbodies with very tenuous attachment to the soil and usually becoming free. Old papery specimens persist for many months and may be blown for considerable distances scattering spores in the process.
Spore-mass: purplish-black.
Habitat: open grassland, dunes. Autumnal, but old weathered specimens may be found at any time.

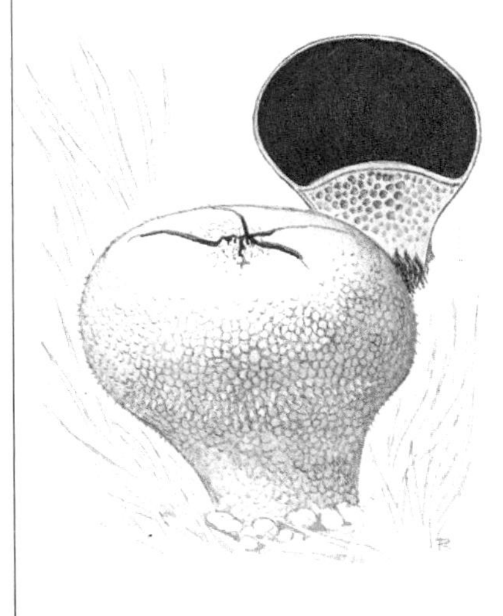

VASCELLUM PRATENSE

Fruitbodies: 2-4 cm diam., pear-shaped with fertile head and sterile stalk, white to cream, ornamented with scurfy white granules and small spines which may be united at the tips. At maturity the spines and granules disappear leaving a more or less smooth, pale brown shiny surface. In this stage a section through the fruitbody shows the fertile head with its dark olive-brown powdery spores held in a cottony mass of threads, separated by a distinct diaphragm from the sterile stalk with its sponge-like texture.
Habitat: in open situations amongst short turf and lawns, often forming fairy-rings. Summer to autumn. Very common.

LYCOPERDON PYRIFORME

Fruitbodies: up to 6 cm high, 3 cm wide, club-shaped or pear-shaped, often narrowly so, pale greyish or pale brownish, densely covered by fine scurfy granules.
Stalk: arising from conspicuous, white, chord-like strands; internal structure sponge-like.
Spore-mass: greenish-yellow then olive-brown.
Habitat: gregarious, often in very large numbers on and around stumps of deciduous trees. Autumnal, but old weathered specimens can be found throughout the year. Very common. Although several other puff-balls grow on decaying leaves in the forests, this is the only one that actually grows on wood.

CALVATIA EXCIPULIFORMIS

Fruitbodies: up to 12 cm high, pestle-shaped with fertile head and well-developed sterile stalk. The surface is pale greyish-buff, densely covered over the upper portion with fine, scurfy whitish hair-like spines, interspersed with minute granular warts. In section the fertile head is seen to contain the powdery olive-brown spores held in a mass of cottony threads, while the sterile base has a whitish sponge-like appearance.
Habitat: pastures, heaths or deciduous woodland. Autumnal. Occasional.

This puff-ball is easily recognized by its tall pestle-like fruit body.

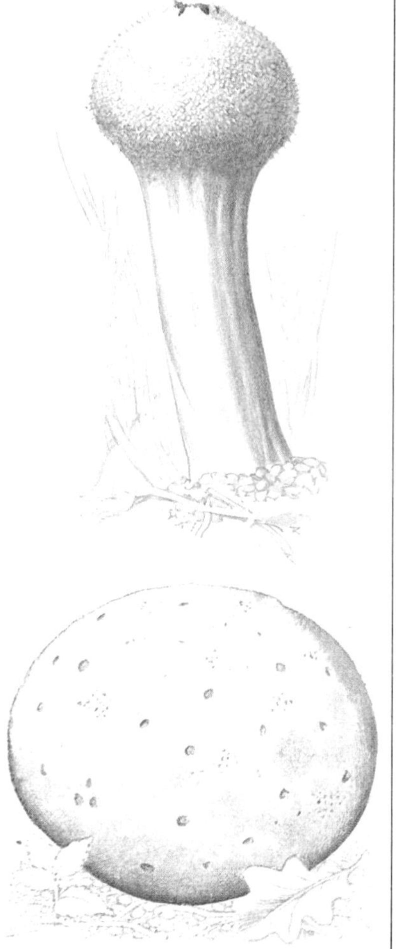

GIANT PUFF-BALL
Langermannia gigantea

Fruitbodies: up to 30 cm diam., occasionally even larger, resembling a smooth white ball with kid-like texture to the surface. The interior is initially white and fleshy becoming yellowish, but as the spores mature the colour changes to olive-brown and the texture becomes cottony. Attachment to the soil is by means of a tiny chord, such that mature fruit-bodies often become free and get blown by wind scattering spores in the process.
Habitat: on the ground, in fields, gardens and woodland. Autumnal. Occasional. Edible if eaten while the flesh is still white.

EARTH BALL
Scleroderma citrinum

Fruitbodies: 5-10 cm diam., hemispherical, often slightly flattened above, with a basal chord-like attachment or sometimes with a mass of cottony threads, yellowish or ochre-coloured with a conspicuously rough, coarsely-scaly surface. When cut through the fruitbody is seen to have a thick whitish rind which often flushes pink, surrounding a firm, purplish-black spore mass which at maturity becomes powdery.
Habitat: sandy heaths or woodland. Summer to autumn. Very common.

STINKHORN
Phallus impudicus

Fruitbodies: 10-14 cm high, exceptionally up to 30 cm, comprising a fragile, white, spongy, hollow stalk with sac-like gelatinous remains of the egg forming a volva at its base, and supporting a pendulous bell-shaped cap at its apex.
Cap: white with reticulate surface which is only visible after removal of the glutinous, olive-coloured spore mass by insects.
Smell: extremely putrid, of rotting meat, often smelled before it is seen.
Habitat: solitary or gregarious in deciduous woodland or gardens, attached by means of white strands to roots or buried wood. Summer to autumn. Very common.

CYATHUS OLLA

This is one of the bird's nest fungi, in which the spores are borne inside little egg-like bodies, which in turn are borne in nest-like cups. The eggs are splashed out by raindrops.
Fruitbodies: 9-15 mm high, 7-13 mm wide, funnel-shaped with broad flaring mouth and narrow base, outer surface felty varying from pale greyish-buff to yellowish-grey-brown or brown; inner surface smooth, shiny, ranging from pale to dark-grey.
Eggs: about 2·5 mm, disc-shaped, either dark-grey or blackish.
Habitat: gregarious, often in large numbers on bare soil in woods and gardens, sometimes in flower pots. Spring to autumn. Occasional.

Ascomycetes

The fungi in this group are extremely varied in shape but all are characterized by bearing their spores in little sacs called asci. The asci may be borne inside the fruitbody or on the surface. The group includes the much sought-after edible truffles and morels, together with many colourful cup fungi and an assortment of smaller species.

CANDLE-SNUFF FUNGUS
Xylaria hypoxylon

This tough, leathery fungus forms strap-like fruitbodies which generally branch like tiny antlers. It gets its name from the powdery tips of the young branches, which resemble snuffed-out candle-wicks. The white powder consists of special spores called conidia. Asci develop later, embedded in small pits nearer the base of the fruit body.
Habitat: tree stumps and other dead wood almost anywhere. All the year.

CRAMP BALLS
Daldinia concentrica

First appearing as rounded, dark brown 'buns' on dead trees, the fruitbodies eventually become jet black and very hard. Up to 9 cm across, they are also known as King Alfred's Cakes. Black spores develop just under the surface, which becomes very sooty when they are released.
Habitat: dead trunks and branches, especially ash trees. All the year.

COMMON MOREL
Morchella esculenta

Cap: 6-8 cm high, yellowish-brown, with sharply angled pits.
Stem: 6-10 cm high, creamy white in colour, almost smooth, but slightly wrinkled at base.
Flesh: fairly thick, whitish, with a delicate flavour. This excellent edible mushroom should never be eaten raw.
Habitat: fields, woods and hedgerows.

TRUFFLE
Tuber aestivum

Fruitbody: 2-6 cm diam. Covered in coarse pyramidal warts. Grey-black. Interior pale greyish-lilac, marbled with darker veins. Asci inside fruitbody.
Habitat: a rare species on chalky soil, growing entirely underground. The related *T. melanosporum* is the famous truffle of France.

ORANGE PEEL FUNGUS
Peziza aurantia

Looking like discarded pieces of orange peel, this rather brittle fungus forms a ragged cup 1-12 cm across. Bright orange inside, where the asci are embedded in the surface, it is downy white on the outside.
Habitat: bare ground or grassy places in woods and many other situations. Late autumn and winter.

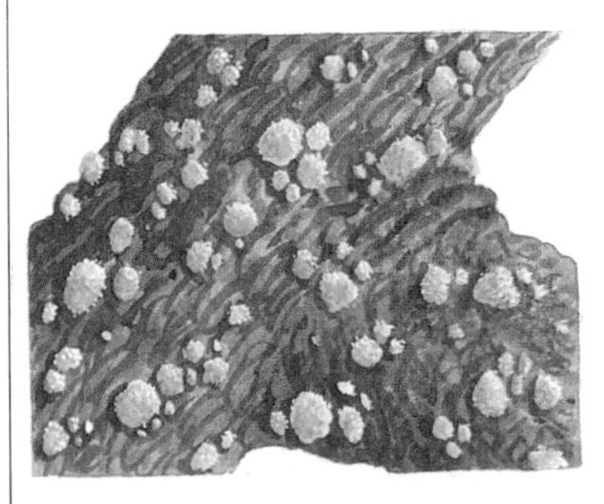

CORAL SPOT FUNGUS
Nectria cinnabarina

This very common lignicolous fungus has 2 visible stages. One consists of many pale pink spots on the wood, each producing numerous non-sexual spores called conidia. The other stage consists of darker red pustules which carry the asci.
Habitat: mainly dead twigs, including garden pea sticks; also on some living twigs, especially after wounding. All year, but asci normally autumnal.

Lichens

Lichens are very hardy plants which can survive in some of the hottest and coldest places on earth. Their bodies consist of fungal threads and minute algae in an intricate partnership. They are tough and dry and quite unlike ordinary fungi, although the fungus partner makes up the bulk of the body. Growth is usually very slow.

There are three main types of lichen. *Crustose* lichens form crusty coats on rocks and other surfaces. They may be broken into hexagonal plates, but they have no obvious lobes. *Foliose* lichens consist of numerous scales or leaf-like lobes, often forming circular patches. *Fruticose* lichens are like miniature bushes, standing erect or hanging in tufts from branches. Many lichens scatter powdery granules, called soredia, which grow directly into new plants. They also reproduce by spores, which are formed in certain regions. The spores are purely fungal and can grow into lichens only if they are quickly joined by the right kind of alga.

DOG LICHEN
Peltigera canina

A large foliose lichen common on sand dunes and in grassy places. Underside is white and felt-like. Lobes are pale grey when dry. Spores are borne in chestnut-brown patches on the edges.

XANTHORIA PARIETINA

One of several similar species, this foliose lichen is abundant on rocks and old walls, on tree trunks and on asbestos roofs. It is especially common by the sea. Spores are formed in darker patches near the centre.

CLADONIA COCCIFERA

One of the 'pixie-cup' lichens, this fruticose species is common on moorland and other areas of peaty soil. The cups grow up from clusters of scales and bear bright red spore-producing patches on their rims.

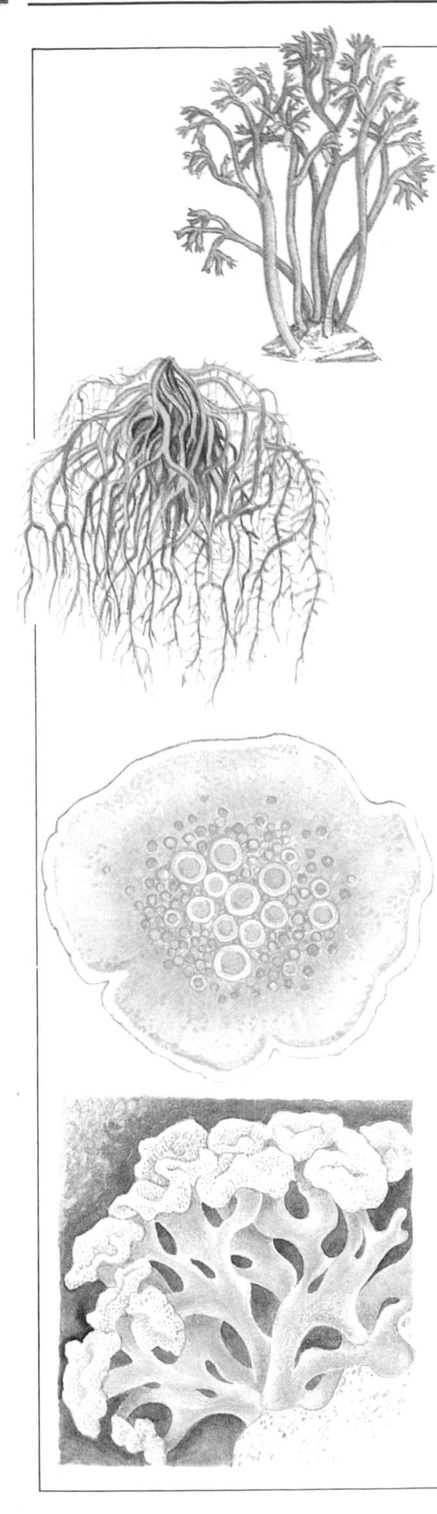

REINDEER MOSS
Cladonia rangiferina

A plant of the Arctic and high moors, this fruticose lichen is the staple diet of the reindeer.

USNEA SUBFLORIDANA

One of the 'beard lichens', this fruticose species forms tangled clumps on tree trunks and branches. Young lichens are usually upright. Soredia are produced in white patches on the stems. The lichen is most abundant in the west, where the climate is damper.

OCHROLECHIA PARELLA

Abundant on old walls and rocks, this is a crustose lichen with no obvious lobes. It forms thick grey crusts as much as 15 cm across and has prominent white margins. Spores are formed in small pinkish brown 'dishes', reminiscent of miniature jam tarts, near the centre of the crust.

HYPOGYMNIA PHYSODES

This is an extremely common foliose lichen of aged tree trunks, rocks and old walls. It forms patches 5 cm or more in diameter and its greyish-green lobes are usually smooth. Soredia are borne under the up-turned tips of the outer lobes. Spores are carried in reddish discs near the centre of the plant, although they are rather rare. The lichen also grows very commonly on old heather stems on heaths and moors, but here the lobes are much paler and they have more distinct brown patches of soredia.

CHAPTER FIVE

SEAWEEDS AND OTHER ALGAE

The algae are simple flowerless plants, most of which live in water. They have no roots and their bodies are never clearly divided into leaf and stem. Many forms, such as *Chlamydomonas* (see page 276), consist of just a single free-swimming microscopic cell. These unicellular algae can multiply very rapidly by splitting into two or more new cells every few hours; their immense numbers are responsible for turning pond and aquarium water green in summer. They also play a vital role in the economy of the sea, for they form the phytoplankton – the 'soup' of minute drifting plants on which the fishes and other marine animals depend for their food. A few unicellular algae live in damp soil, and also on tree trunks which they turn green in very wet conditions.

Seaweed Groups

There are several different groups of algae, not all closely related, and they are usually classified according to their colour – which depends on the types of pigments in their cells. Seaweeds are the largest and most familiar of the algae and they belong to three main groups: green, brown and red. They have a much more complex structure than the other types of algae and some are several metres long. They are often attached to the rocks by sucker-like discs called holdfasts; there may be a stalk-like region between the holdfast and the flat blade, but there are never any special tubes for carrying water and food, such as are found in higher plants. The detailed identification of the larger seaweeds relies mainly on shape and colour and is not usually difficult.

Although many seaweed species are brown or red, all seaweeds contain green chlorophyll and make food by photosynthesis in the same way as other plants. The brown and red pigments that mask the chlorophyll improve the absorption of light under the water. Red is the best light-absorber under water, so red seaweeds can grow in deeper water than other seaweed types – though none can grow in very deep water where there is no light at all. Most seaweeds grow around the coast and are often ripped up and thrown on to the shore during storms.

Green Algae

Green algae (Chlorophyceae) live in fresh and salt water and a few live on land. Many are single-celled organisms, while many more exist as fine threads. Green seaweeds live in the shallowest waters around the shore and also cover the mud of many estuaries. They can withstand considerable exposure to the air. A coating of mucus ensures that they do not dry up too quickly, and also makes them slippery to walk on.

CHLAMYDOMONAS

A genus of microscopic unicellular algae, abundant in fresh water. The plant swims by waving its flagella. The red spot detects light and helps the alga to orientate itself in the best position for absorbing light into the cup-shaped chloroplast where food is made.

SPIROGYRA

A genus of filamentous green algae with several common species in ponds and streams – either floating or fixed to stones: commonly known as blanket weeds. Green chloroplasts run spirally round in each cell. Division of the cells leads to increased length of the filaments. Asexual reproduction takes place by fragmentation, each broken piece growing as a separate plant.

SEA LETTUCE
Ulva lactuca

A world-wide plant with fronds like very thin lettuce leaves, up to 45 cm long. Found on all shore zones, but mainly in the upper parts; especially common in muddy estuaries and bays and wherever fresh water runs over the shore. Edible.

ENTEROMORPHA INTESTINALIS

Sometimes known as grass kelp, this very common seaweed grows on rocks and mud in estuaries and on the upper shore zones, especially where streams trickle over the shore. The thallus, up to 60 cm long, is tubular, rather like a sausage skin inflated with gas.

Red Algae

The red algae (Rhodophyceae) are almost all marine and they are most common in the warmer seas. Some live in rock pools, but the majority inhabit deeper waters. They cannot survive much exposure to the air. The colour ranges from pale pink to violet or brownish-red and the range of form is tremendous. There are a few unicellular species; some encrust the rocks like lichens; many are filamentous; and others have leaf-like blades. The fronds are often elaborately branched. Many secrete coatings of calcium carbonate around themselves and are sometimes mistaken for corals.

CORAL WEED
Corallina officinalis

A stiff, bushy seaweed coated with lime and reaching 15 cm. Pink to purple in life; white when dead. It grows in shady pools and on stones from the middle shore down to fairly deep water. It can survive some exposure to the air in shady places.

RED LAVER
Porphyra umbilicalis

Has very thin, leaf-like lobes up to 25 cm across; often split into ribbon-like strips. Ranges from purplish-red to olive-green; usually green when young. Grows on rocks and stones on all parts of the shore, especially where there is sand as well. It is eaten in many places.

IRISH MOSS
Chondrus crispus

Also known as carrageen, this common seaweed grows in rock pools and on other rocks on the middle and lower shore. The fronds reach 15 cm and range from pink to deep reddish-brown; sometimes green in sunlit pools. Frond divisions may be much narrower in wave-battered areas. The plant is eaten in many places.

POLYSIPHONIA NIGRESCENS

The finely branched, feathery fronds reach 15 cm. It grows on rocks on the lower shore and below low-water mark, and may also be found growing on the larger brown seaweeds.

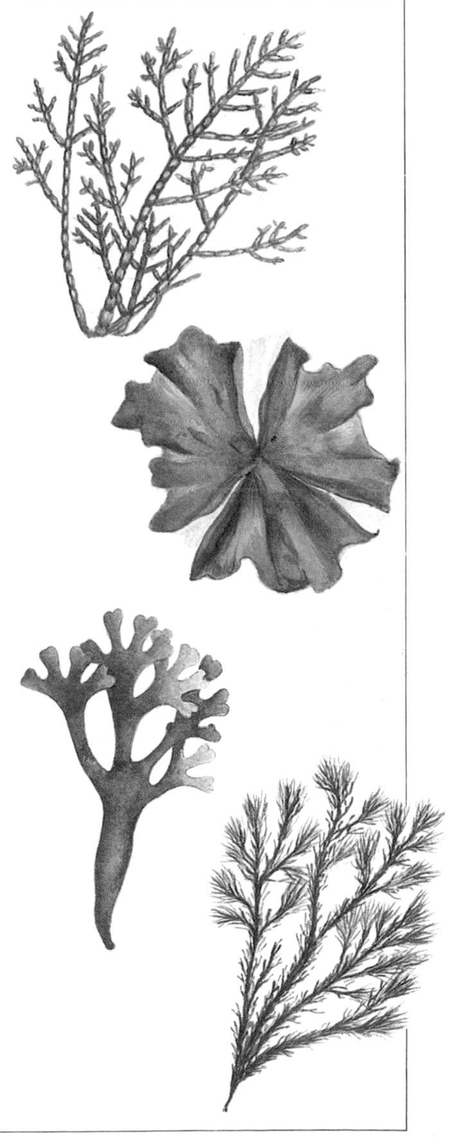

Brown Algae

Nearly all the brown algae (Phaeophyceae) are marine and they include the largest seaweeds. None is unicellular. They are the commonest seaweeds of cool regions, occupying zones from the middle shore down to below low-tide level. They often form dense carpets on the rocks. This is especially true of the tough-fronded wracks, which can withstand a good deal of battering by the waves.

BLADDER WRACK
Fucus vesiculosus

Also known as popweed because of the air-filled bladders which buoy up the fronds in the water, this very tough seaweed occupies large areas of the middle shore. Fronds reach 90 cm. It often has no bladders where pounded by very strong waves. The pale swollen tips of the fronds house the reproductive organs.

SERRATED WRACK
Fucus serratus

Easily recognized by its serrated margins, this seaweed grows abundantly on the middle and lower shore – usually just below the bladder wrack zone; it avoids the most exposed areas where wave action is strongest. The fronds reach 1 m or more and are quite flat; swollen tips are much less obvious than in other wracks.

THONGWEED
Himanthalia elongata

Grows on rocks on the middle and lower shore, usually just below the serrated wrack. The strap-like fronds may reach 3 m. Young plants resemble buttons until the thongs begin to grow.

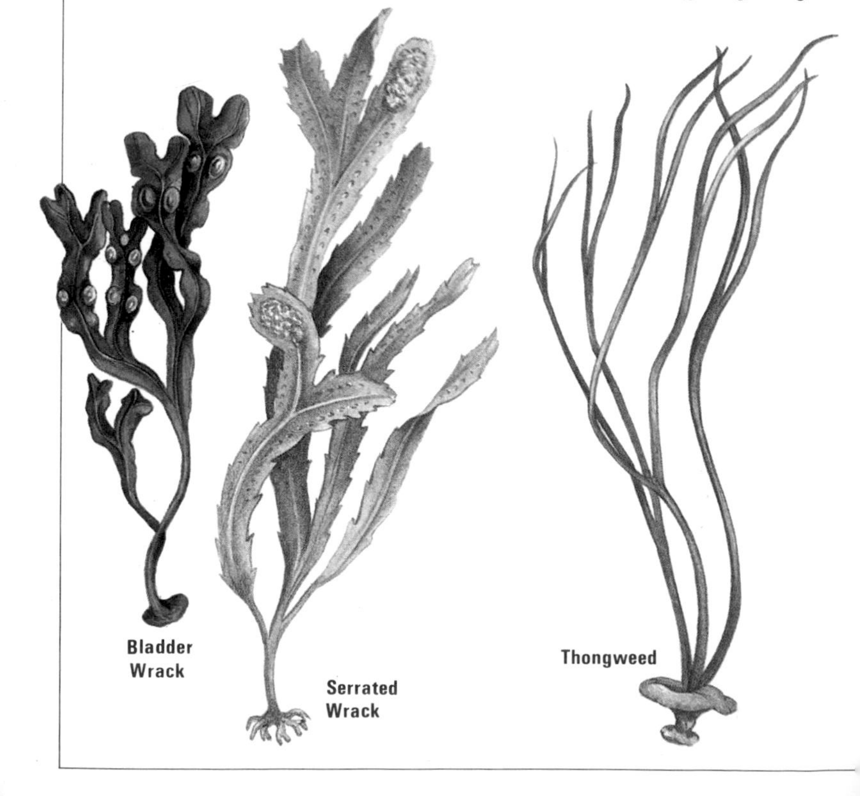

PEACOCK'S TAIL
Padina pavonia

The curved, fan-shaped frond is up to 15 cm high. It grows on rocks and stones in rock pools and around low-water mark. Common on Mediterranean and south-western coasts of Europe: reaches Britain, but uncommon and found only in summer and autumn – mainly on sunny sides of rock pools.

CHANNELLED WRACK
Pelvetia canaliculata

Named for the distinct channels along the branches, this seaweed is abundant on the upper zones of rocky shores and also on estuarine mud-flats and salt-marshes. It can survive out of water for several days at a time, relying on water trapped in its narrow channels. The fronds reach 15 cm.

OARWEED
Laminaria digitata

Also known as tangle, this seaweed forms dense beds from about low-water mark down to 30 m. A branched holdfast grips the rocks and the stalked, rubbery frond is up to 3 m long. It is split into numerous ribbon-like strips. When uncovered by the tide it lies flat: the stalks of the very similar species (*L. hyperborea*) remain upright.

English Names

Scientific Names

Most of the societies and organizations given here cater for amateur as well as professional interests. Many have libraries that members can use and most run schemes to help with the identification of material. Journals or newsletters for members are produced regularly by many of the societies.

Botanical Society of the British Isles
c/o British Museum, Natural History, Cromwell Road, London, SW7 5BD

Arranges lectures, exhibitions and field meetings all over the country, enabling amateurs to meet professionals and to learn about flowering plants and ferns.

Botanical Society of Edinburgh
c/o Royal Botanic Gardens, Inverleith Row, Edinburgh, EH3 5LR

Devoted to the study and protection of wild flowers through lectures and excursions. Strong emphasis on young people.

British Bryological Society
c/o Dept of Botany, The University, Whiteknights, Reading, RG6 2AS

Devoted to the study, identification and conservation of mosses and liverworts.

British Lichen Society
c/o British Museum, Natural History, Cromwell Road, London, SW7 5BD

Meetings, excursions and journals.

British Mycological Society
c/o Royal Botanic Gardens, Kew, London, TW9 3AB

Devoted to the study of fungi.

British Naturalists' Association
6 Chancery Place, The Green, Writtle, Essex, CM1 3DY

Promotes interest in all branches of natural history, through local and national meetings and regular publications.

British Pteridological Society
c/o Botany Department, British Museum, Natural History, Cromwell Road, London, SW7 5BD

Devoted to the study of all aspects of ferns and other pteridophytes.

Field Studies Council
62 Wilson Street, London, EC2A 2BU

Promotes a better understanding of our environment. Runs field courses in all branches of natural history and countryside appreciation at ten centres in England and Wales.

Royal Society for Nature Conservation
The Green, Nettleham, Lincs. LN2 2NR

Co-ordinates all the County Naturalists' Trusts, of which there are more than 40 in Britain. Manages many important reserves.

Scottish Field Studies Association
Kindrogan Field Centre, Enochdhu, Blairgowrie, Perthshire, PH10 7PG

Encourages the study and conservation of wildlife and the countryside in Scotland. Organises numerous courses on a wide variety of wildlife topics.

Urban Wildlife Group
11 Albert Street, Birmingham, B4 7UA

Promotes the study and conservation of all forms of wildlife associated with the urban environment.

Wild Flower Society
68 Outwoods Road, Loughborough, Leics.

Devoted to the study of wild flowers.